The Boys' Book of Engine-Building: How to Make Steam, Hot Air and Gas Engines and How They Work, Told in Simple Language and by Clear Pictures

Archie Frederick Collins

This work has been selected by scholars as being culturally important, and is part of the knowledge base of civilization as we know it. This work was reproduced from the original artifact, and remains as true to the original work as possible. Therefore, you will see the original copyright references, library stamps (as most of these works have been housed in our most important libraries around the world), and other notations in the work.

This work is in the public domain in the United States of America, and possibly other nations. Within the United States, you may freely copy and distribute this work, as no entity (individual or corporate) has a copyright on the body of the work.

As a reproduction of a historical artifact, this work may contain missing or blurred pages, poor pictures, errant marks, etc. Scholars believe, and we concur, that this work is important enough to be preserved, reproduced, and made generally available to the public. We appreciate your support of the preservation process, and thank you for being an important part of keeping this knowledge alive and relevant.

THE BOYS' BOOK
OF ENGINE-BUILDING

A Model Locomotive built by a famous American Engineer when a boy

THE BOYS' BOOK OF ENGINE-BUILDING

HOW TO MAKE STEAM, HOT AIR AND GAS ENGINES AND HOW THEY WORK, TOLD IN SIMPLE LANGUAGE AND BY CLEAR PICTURES

BY

A. FREDERICK COLLINS

Author of "Design and Construction of Induction Coils," "Manual of Wireless," "Keeping Up with Your Motor Car," "How to Fly," "The Book of Wireless," "Shooting for Boys," "Inventing for Boys," etc.

With Drawings by the Author

BOSTON
SMALL, MAYNARD AND COMPANY
PUBLISHERS

Copyright, 1918
By SMALL, MAYNARD & COMPANY
(INCORPORATED)

A WORD TO YOU

I hope you will read these few pages about a boy who built model engines along in the late 70's out in the Far West and at a time when and a place where it was next to impossible to get materials for his work.

This boy was Bion J. Arnold, now our foremost American engineer. He was only thirteen years old when he built his first engine—and it worked, too. It was a little horizontal steam affair about seven inches long, and he went about building it in the right way—that is he made patterns for the chief parts, cast them in lead and put them together with the tools of a kindly disposed gunsmith.

When Bion was fourteen, he built a vertical steam engine over a foot high and here is where his genius showed itself again, for he used a piece of old iron pipe for the boiler, a discarded hub from a wagon wheel for the fire-box, a wheel from a valve, which had been thrown away, for

A Word to You

the flywheel, a gas-cock some one had given him for the throttle valve and, finally, he riveted the fire-box to the boiler with bolts which he had forged with his own hands.

He built his third engine when he was fifteen; the main parts of this he cast in Babbitt metal and he made the boiler of ⅛ inch thick sheet iron which he hammered into shape and riveted together at the forge of a genial blacksmith. The heads of this boiler were made of two wooden disks and these were *tied* in place by passing four iron rods through them and held on with a couple of nuts.

His next attempt was made when he was sixteen years old. His idea was to build a 2 horse power engine to drive the wood saw with—a noble scheme since it was his job to keep enough wood in the box for the kitchen fire—but his father would have none of either the scheme or the engine and so both were scraped somewhere between the time it was concocted and the place where it would have been put to work.

Having been so successful in building these small model engines young Arnold began his last model when he was seventeen, and this was a

A Word to You

miniature locomotive of the type that was in common use on western railroads in 1876, and for about ten years thereafter. A picture of his handiwork is shown in the frontispiece of this book.

He took the measurements for it from the big locomotives in the roundhouse at Lincoln, Nebraska, where he was going to school and, with his father financing his hobby, after many months of hard work he finished his model engine.

It is complete down to the smallest details, if we except the air brakes and the air compressor for it, but even these were made though he never put them on. This beautiful little model—it is three feet long—can be seen in Mr. Arnold's Chicago office where the master builder now conceives and directs various kinds of engineering work on a gigantic scale.

When the model locomotive was fired up with real coal and a few pounds of steam were raised in the boiler, it ran as smoothly as any of its big prototypes, much to the pleasure of the builder and the delight of those who saw it in operation.

Now the point is this: To build a working engine however small takes a certain amount of

A Word to You

mechanical ability plus stick-to-it-ive-ness, and if you have these qualities there isn't a reason in the world why you should not become a great engineer—not so great, perhaps, as Mr. Arnold, for his income now as a traction engineer is considerably more than that of the President of the United States.

If it is your intention to become a civil, mechanical or an electrical engineer—Mr. Arnold is all three—the thing for you to do is to learn how to use tools as well as a mechanic does and then you want to get the best technical schooling you can.

With this equipment, provided you have a fair amount of business tact too, you will be assured of a large income as long as you live. And this is why you should start in now and build engines.

A. FREDERICK COLLINS.

550 Riverside Drive,
  New York City.

CONTENTS

CHAPTER	PAGE
I.—The First Engines	1
II.—Two Simple Steam Turbine Engines	24
III.—A Simple Piston Steam Engine	45
IV.—A 1/24-H. P. Horizontal Steam Engine	61
V.—Making Small Boilers	91
VI.—Fittings for Model Engines	110
VII.—A Model Atlantic Type Locomotive	138
VIII.—A Model Atlantic Type Locomotive (Con.)	156
IX.—Steam, the Giant Power	182
X.—A Hot Air, or Caloric, Engine	199
XI.—A 1/8-H. P. Gas Engine	214
XII.—Useful Information	241
Appendices	247

THE BOYS' BOOK OF ENGINE-BUILDING

CHAPTER I

THE FIRST ENGINES

Hero Invents the Steam Engine—Branca Devises a New Engine—Papin Gets up the Piston Engine—Newcomen Makes the Engine Work—The Boy Who Made it Self Acting—Watt Puts on the Flywheel—The Development of the Steam Boiler—The First Steamboats—The Invention of the Locomotive—The Modern Steam Turbine—Some Other Kinds of Engines: The Compressed Air Engine; The Hot Air Engine; The Gas Engine.

About the beginning of the Christian era, a babe named Hero first saw the light of day in far off Alexandria in Egypt, and this child was destined to make a lasting impression on the human race.

Even as a boy he built apparatus and machines of all kinds, even as you do now, but this was doubly hard in his day for few inventions had been made and tools to make them with were just about as scarce, for science had not yet come into its own.

Hero Invents the Steam Engine.—But Hero was not discouraged and he began to invent and make things for himself. To tell you of all the experiments he made and all the apparatus he built would take a whole book of this size, but his greatest invention and the one that interests us now is his steam engine, for he was the first who ever made steam do useful work.

Hero called the steam engine which he invented and built an *eolipile* (pronounced *a-ol'-i-pil*) and this curious word he coined from two Latin words, the first of which is *Eolus,* the name of the *god of the winds,* and the other one *pila,* which means *ball.*

Let's find out now just why Hero named his engine the god of the winds plus a ball? In the first place he made his engine of a hollow metal ball and he fitted two short bent pipes into holes on the opposite sides of it. Next he mounted the ball on a pair of hollow pipes, or *trunnions* as they are called, so that it would revolve easily and the lower ends of these pipes were fixed to a boiler, all of which is shown in Fig. 1.

When he built a fire under the boiler the steam passed through the trunnion pipes into the ball and then out of the bent pipes into the air. As the steam struck the air with considerable force

Fig. 1. Hero's Reaction Engine or Eolipile

it acted much the same as a rubber ball does when it strikes a sidewalk, that is it *reacted* on the hollow ball forcing it to turn in the opposite direction and at a fairly high speed, like unto the winds let loose by the god Eolus. This is the reason an engine of this kind is called a *re-action engine*.

Branca Devises a New Engine.—Strangely enough nothing more worth while was done in the way of transforming the energy of steam into mechanical motion during all the long centuries from the year 1 to 1609.

Then an Italian named Branca devised another sort of rotary steam engine. He made a paddle-wheel and when the steam from a boiler was directed through a spout, or more properly a

Fig. 2. Branca's Impulse Engine or Turbine

nozzle, against the paddles, or *blades,* as shown in Fig. 2, the force of the *impulses* of the steam striking them made the wheel revolve. For this reason an engine of this kind is called an *impulse* engine.

In those early days the Italians put an artistic touch on everything they made—indeed they haven't got over the habit yet—and so it was perfectly natural for Branca to give his boiler the head and body of a human being and he added to the life-like effect by having the steam blown from its mouth.

Now you may think out loud that a reaction ball like Hero's or an impulse wheel like Branca's is not a real steam engine but if you do you are quite mistaken for it so happened that within the last twenty years both of these ancient toys have led to the last word in steam engine building and that is—the *steam turbine*.

Papin Gets Up the Piston Engine.—The first steam engine in which steam in a cylinder moved a piston forth and back, or up and down, came shortly before Branca devised his rotary engine but these engines were used only for pumping water, for no one knew how to change the *reciprocating motion*, as the to and fro motion of the piston is called, into *rotary motion*, as the complete turning of a wheel is called, for another hundred years or so.

To begin at the beginning of the first invention of the piston engine, we have to start with Dennis Papin, a Frenchman. In 1610 he got up the scheme of using a piston in a cylinder and moving it by steam; I say *scheme* because it did not work successfully when he built it. The trouble with his engine was that he made the boiler and the cylinder in one piece.

The way he tried to use it was to pour a little water in the bottom of the cylinder, which stood in a vertical position. Then he heated it as shown in Fig. 3. When steam was made it

Fig. 3. Papin's; The First Piston Engine

forced the piston up, the fire was then taken away, the steam cooled down, thus forming a *vacuum* in the cylinder, and the *pressure of the air* on the piston forced it down again.

Although this engine of Papin's was intended to work half by steam and half by air pressure it was called an *atmospheric engine* and this is the name it has always been known by.

The First Engines

Newcomen Makes the Engine Work.— Nearly a hundred years later, in 1705, Thomas Newcomen, of England, and his assistant, James Cawley, made the piston engine of Papin work. They accomplished this by using a boiler that

Fig. 4. Newcomen's Improved Engine

was separate and distinct from the cylinder. The piston in the cylinder was connected by a rod to one end of a lever called a *walking beam* and a weight to balance it was hung on the other end of the beam as shown in Fig. 4.

When steam was admitted into the cylinder from the boiler, it forced the piston up, then the steam *valve* was shut off and a little cold water was let into the cylinder to *condense* the steam, that is to change it into water again, when a vacuum would be formed; this done, the piston was pushed down by the pressure of the air on it; when the steam was turned into the cylinder again, it forced the water out through another valve.

The end of the walking beam which carried the weight was connected to a pump and hence the Newcomen engine came to be largely used in England for pumping water out of mines.

The Boy Who Made It Self-Acting.—As you can imagine the piston of Newcomen's engine worked very slowly and the steam and water valves had to be opened and closed by hand. As it was light, though tedious, work to do this, a boy was usually given the job and a youngster named Humphrey Potter happened to be one of them.

Now Humphrey was a boy of genius, and genius and humdrum work are seldom on speaking terms. So he bethought him to rig up some levers worked by strings tied to the walking beam to open and close the valves and his inven-

The First Engines 9

tion worked like a charm. He was in a fair way, at least it seemed so to him, to draw his sixpence a day while the engine worked automatically.

Instead his genius threw himself and all the other valve boys in England out of their jobs, and all he ever got out of it was undying fame.

Fig. 5. Potter's Self-Acting Engine

Humphrey's invention was at once improved upon by one Henry Beighton who connected a rod to the walking beam to work the valves and it is quite likely that, being a man, he profited by it.

Watt Puts on the Flywheel.—In your young

lifetime you have heard enough of James Watt, of Glasgow, Scotland, to believe, probably, that he was the inventor of the steam engine, but you have seen that Papin and Newcomen and Potter

Fig. 6. Watt's; The First Real Engine

developed it to a considerable extent before Watt took it up.

What Watt did, though, constituted a mighty invention for it was he who made the wheels go round. It was in 1763 that he conceived the idea of connecting the piston rod to a crank having a flywheel fixed to it, and this made the engine useful for producing power for a thousand different purposes where before it was limited to pumping water. It is shown in Fig. 6.

To make the piston turn a crank the steam had to act on it much faster than it did in the Newcomen engine, so Watt let the steam escape from the cylinder at the end of each stroke, instead of condensing it in the cylinder by letting water run into it. Not only this, but by keeping the cylinder hot the engine used far less steam, and so it was much more economical.

Some years after Watt made two other improvements of great value, first of which was to admit steam on each side of the piston alternately, thus making the engine *double acting;* and the second was to cut off the supply of steam to the cylinder before the piston had completed its stroke and so using the *expansive power* of the steam in the cylinder to force the piston on to the end of its stroke.

Finally, Watt invented the *throttle valve,* so that the amount of steam which flowed into the cylinder could be controlled; and he also invented the *centrifugal governor* which opens and closes the throttle valve automatically, depending on the speed with which it revolves and this in turn regulates the speed of the engine.

The Development of the Steam Boiler.—Since the first engine was the one built by Hero, we are bound to believe that the first steam boiler was the one which he used to generate steam for it. His boiler, as shown in Fig. 1, was simply a hemispherical shell set on a stand and heated by a fire under it.

In the early days of the piston engine boilers were built in the shape of long boxes with fires under their entire length, but they were so weak that a steam pressure of only three or four pounds to the square inch could be used with safety. For this reason cylindrical boilers, which were very much stronger, took the place of the *rectangular* boilers.

The greatest improvement in boiler making came, however, when some genius put a *central flue* in it and fixed the open ends to the boiler heads, or *sheets,* as they are called, so that there is a hole clear through the middle of the boiler.

This is surrounded by water and the heat of the fire passes through it, as shown in Fig. 7.

As this scheme proved so successful, a large

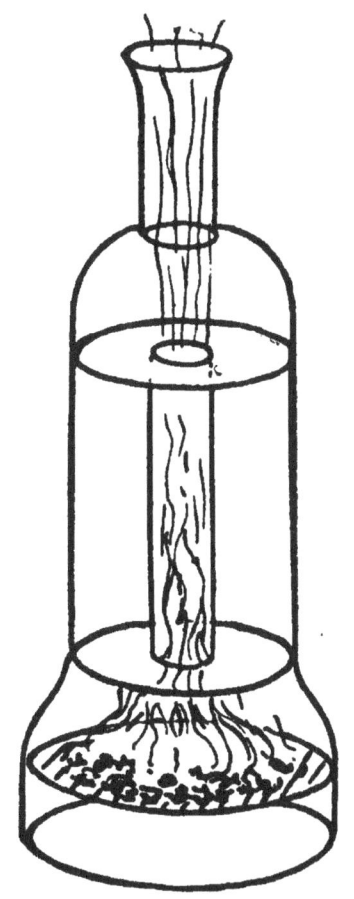

Fig. 7. The Fire-Tube Boiler

number of *fire tubes,* as these flues are called, were then used for they gave a larger *heating surface* and, hence, a higher pressure could be kept up. The tubular boiler, as this type is known, see Fig. 8, is largely used for stationary engines and is the only kind used for locomotives.

The First Steamboats.—Long before a steam engine capable of doing useful work had been invented, men began to think about using steam to propel boats.

Fig. 8. A Modern Tubular Boiler

The first boat actually driven by a steam engine was a *packet* built by John Fitch, of England, in 1787. It was a pretty crude attempt, for it used paddles at the sides and these were moved with a feathering stroke very like that used when you paddle a canoe. It ran at a speed of about three miles an hour.

Fitch built another steamboat moved by paddles at the stern, but his third boat was a great im-

The First Engines 15

provement over the other two, for it was driven by a small propeller. From 1785 to 1788 Miller and Symington, also of England, experimented with a steam propelled pleasure boat, but their efforts did not lead to any practical results.

In 1803 Robert Fulton, of your own United States, was in Paris and while there built a small

Fig. 9. Fulton's Steamboat the *Clermont*

steamboat and operated it on the Seine with great success. Coming to America he built the *Clermont,* in 1807, a small paddle-wheel packet, a picture of which is shown in Fig. 9, and launched her on the Hudson River.

When she started on her trial trip from New York to Albany, thousands of eager spectators lined the shore and marveled at the ease and speed with which she sailed up the river. She reached Albany without mishap and thus it was the great era of steamboats and steamships began.

The Invention of the Locomotive.—The first locomotive built was really a steam automobile, that is it was made to run on ordinary roads.

Since the engines were poor and the roads were bad, Richard Trevithick, of England, built a locomotive in 1804 to run on a track. Being fearful that the wheels would slip on the smooth surface of the rails he used a cog wheel that *meshed* with a cog rail.

Matthew Murray, also of England, built the next locomotive in 1811. It had two upright cylinders set on top of the boiler; the piston rods were connected with a crankshaft, and a wheel on this turned the driving wheels. The first engine to use a fire tube boiler was made in 1813 by William Hedley, likewise of England, and his engine was also the first to make the spent steam from the cylinders exhaust through the smokestack to give the furnace a good draft.

It was George Stephenson—he was of England too—who put the finishing touches on the loco-

The First Engines 17

motive in 1814. This he did by connecting the piston rods directly to the crank pins on the driving wheels and he linked the front and back driving wheels with a *coupling rod*.

Stephenson built the *Rocket* in 1829, see Fig.

Fig. 10. Stephenson's Locomotive, The Rocket

10, and in a test on the Liverpool and Manchester Railway it ran away from its competitors, all of which were driven by sprocket wheels and chains. He not only won the prize, but fixed the design for locomotives from that day to this.

The Modern Steam Turbine.—And now we come back to the place where we started from

2000 years ago, which goes to prove the truth of the saying, *"The first shall be last."* Engineers of our own time have perfected the reaction engine of Hero and the impulse engine of Branca and by combining the principles of both they have made an engine which uses far less steam and is much smaller for the horse power developed than the best reciprocating engines now in use. It is not called a steam engine, however, but a *steam turbine* if you please, though it is an engine just the same.

Though other men had built steam turbines, it was Gustav De Laval, of Sweden, who, in 1883, made it a practical success. He did this by driving a cream separator with it. His turbine is a highly developed Branca's impulse engine having a single wheel mounted on a flexible shaft. The small ones run at the enormously high speed of 30,000 revolutions per minute and this is slowed down to working speeds by means of what are called *reduction gears*. It is shown in Fig. 11.

A year after De Laval got out his single wheel turbine, Charles A. Parsons, of England, combined the principles of the reaction and impulse

The First Engines

engines and built the first *multiple steam turbine*. This turbine, see Fig. 12, developed 10 horse power when running at 18,000 revolutions per

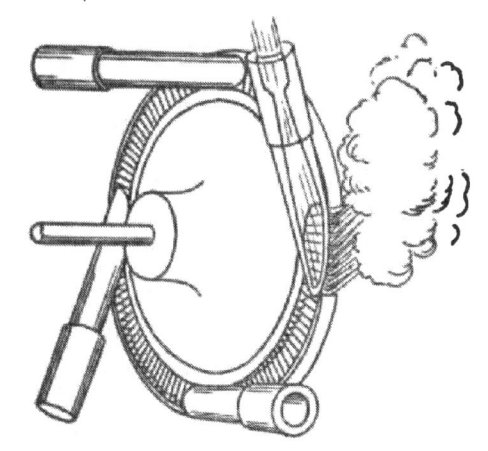

Fig. 11. De Laval's, the First Practical Turbine

minute and its success led to the building of steam turbines powerful enough to drive the

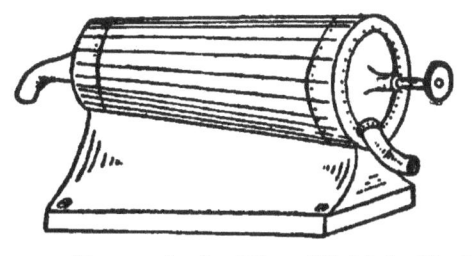

Fig. 12. Parsons', the First Multiple Turbine

largest steamships across the Atlantic Ocean in less than five days.

Other Kinds of Engines.—There are several

other ways to run engines besides using steam. Among them are by (1) *compressed air;* (2) *hot air* and (3) *gas,* or *gasoline*, which amounts to the same thing.

The Compressed Air Engine.—Any kind of an engine that can be run by steam can be run by compressed air. A simple way to compress air for running a small engine is described in Chapter V.

Compressed air is used instead of steam for running large engines when (a) the source of power must be a long way off from the engine; (b) when the smoke is objectionable as on street cars and in submarines, and (c) where the heat from the fire-box is dangerous, as in mines.

The Hot Air Engine.—The *hot-air,* or *caloric engine* as it is sometimes called, is so built that the heat of the fire is used to produce mechanical motion by acting directly on the piston.

This kind of an engine was invented by Robert Stirling, of England, in 1816 and in 1827 he built his first practical engine. John Ericsson, of Sweden, made an improved hot-air engine in 1833, but it was not until 1852 that he built his first commercial engine, see Fig. 13, and engines

The First Engines

of this type are largely in use at the present time for pumping water.

The Gas Engine.—In this engine a fuel gas, such as coal gas, or a gas formed of gasoline, mixed with air, is drawn into the cylinder and,

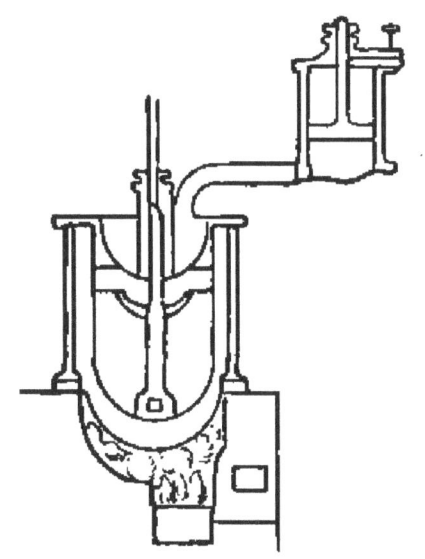

Fig. 13. Ericsson's Hot Air Engine

after it is compressed by the piston, is exploded usually by a hot tube, or an electric spark, which produces what is called the *power stroke*.

The beginning of the gas engine dates back to 1678 when the Abbé d'Hautefeuille, of France, conceived the idea of using the explosive power of gunpowder on a piston to make it do work.

22 *The Boys' Book of Engine-Building*

Philip Lebon, also of France, patented an engine in which he proposed to use coal gas as the explosive fuel.

The first gas engine that actually worked was invented by Lenoir, likewise of France, in 1860. He used gas for the fuel and this was exploded

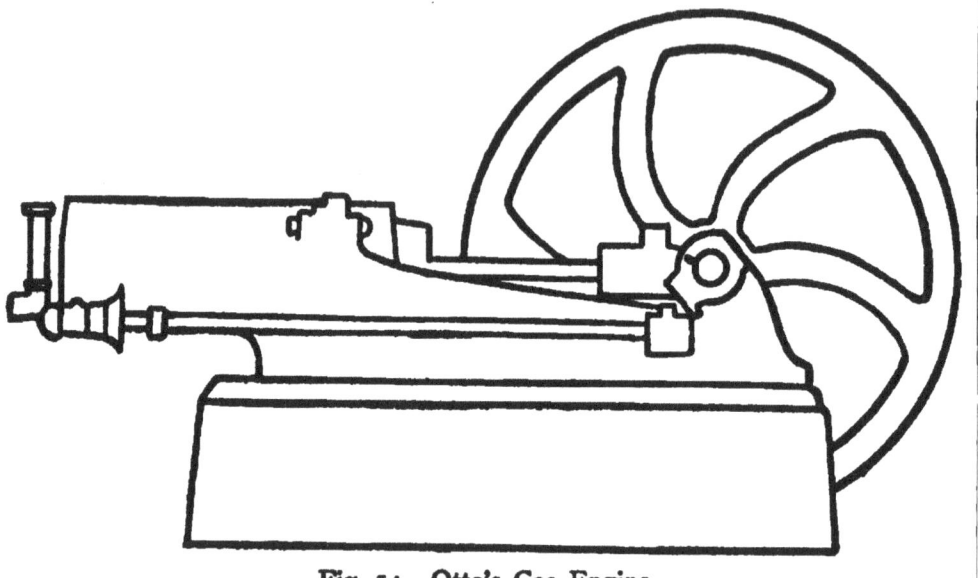

Fig. 14. Otto's Gas Engine

by an electric spark, but since the gas was exploded without being compressed, his engine was very wasteful and inefficient.

To Dr. Otto, of Germany, belongs the credit of having brought the gas engine to a successful conclusion in 1878. This was done by drawing in the gas on the *suction stroke,* compressing it on the *compression stroke,* exploding it to produce the *power stroke* and exhausting the burnt gases on the *exhaust stroke.* Hence this kind of a gas engine is known as the Otto *four cycle* engine and to its development is due not only the automobile, but that mightier marvel, the airplane.

CHAPTER II

TWO SIMPLE STEAM TURBINE ENGINES

Making the Boiler—Forming the Turbine Wheel—An Easily Made Alcohol Lamp—Running the Turbine—A Model De Laval Turbine—The Tools Required—Making the Turbine Wheel—Forming the Blades—Fashioning the Nozzle—The Turbine Wheel Case—The Bearings for the Shaft—Mounting the Wheel—The Reduction Gears—How the Steam Turbine Works: The Action of the Blades; How the Nozzle Works.

A Simple Steam Turbine Engine.—Just as hundreds of other fellows before you, from Branca to the present time, have built impulse steam engines for their first models, so you would do well to follow in their footsteps and make one too.

To make a toy paddlewheel engine, or *turbine*, as it is now called, is a matter of a couple of hours' work and it will give you many times that many hours' pleasure, for steam in action is a wonderful power and what is equally to the point it has done more to make our world a cheerful place to live in than any other power harnessed by man, not even excepting electricity.

Now there are two chief parts to every steam power plant, however small or large as you have had occasion to observe in the first chapter, and there are, (1) the *boiler* which generates the steam and (2) the *engine* which transforms the energy of the steam into *rotary mechanical motion,* which means in everyday English that the force of it turns a wheel around.

Making the Boiler.—To make this little steam turbine get a ¼ pound baking-powder can—empty of course—and soak it in water until the paper label comes off; this is for the boiler.

Drill or punch a ⅛-inch hole in the can for the nozzle, about 1¼ inches back from the open end and drill or punch four 1/16-inch holes, two on each side, at the places shown in the side and end views in Figs. 15 and 16 for the rivets; and, finally, cut a ⅜-inch hole in the cover for the *filler,* that is the hole through which water is poured into the boiler.

For the nozzle get a ⅛-inch piece of brass or copper tube, if possible, ½ inch long, or make one of tin, if necessary, and flatten one end a little.

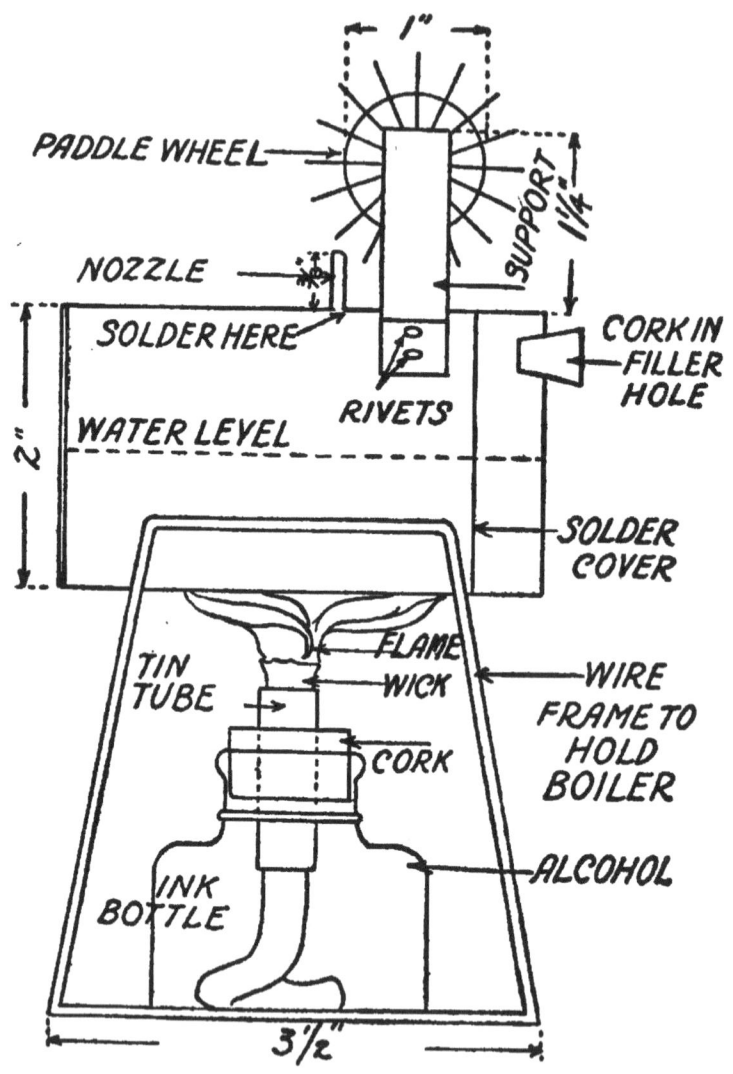

Fig. 15. Side View of Turbine

Now solder the nozzle over the ⅛-inch hole in the can and don't be afraid of using too much solder.

Cut out two strips of sheet brass, copper or heavy tin ½ inch wide and 2 inches long for the supports that hold the wheel, and rivet these to the can so that they will be ⅝-inch apart. You can buy small rivets or make them yourself by cutting off bits of copper wire. When you have these things done, solder the cover on tight.

Forming the Turbine Wheel.—Next make the turbine wheel. Get, or saw out, a wood wheel ½ inch thick and about 1 inch in diameter, or if you can get a spool, on which ribbon is wound, of the right size it will serve the purpose very well.

Saw 16 slots ⅛ inch deep at equal distances apart around the *periphery,* that is the rim of the wheel, and cut out as many pieces of tin ½ inch wide and ½ inch long for the paddles, or *blades,* or *buckets,* as these members are variously called. Fix the blades into the slots, using sealing wax on the edges, if needs be, to keep them in tight.

Use a piece of perfectly straight iron or brass wire, or a wire nail will do, ⅛ inch in diameter

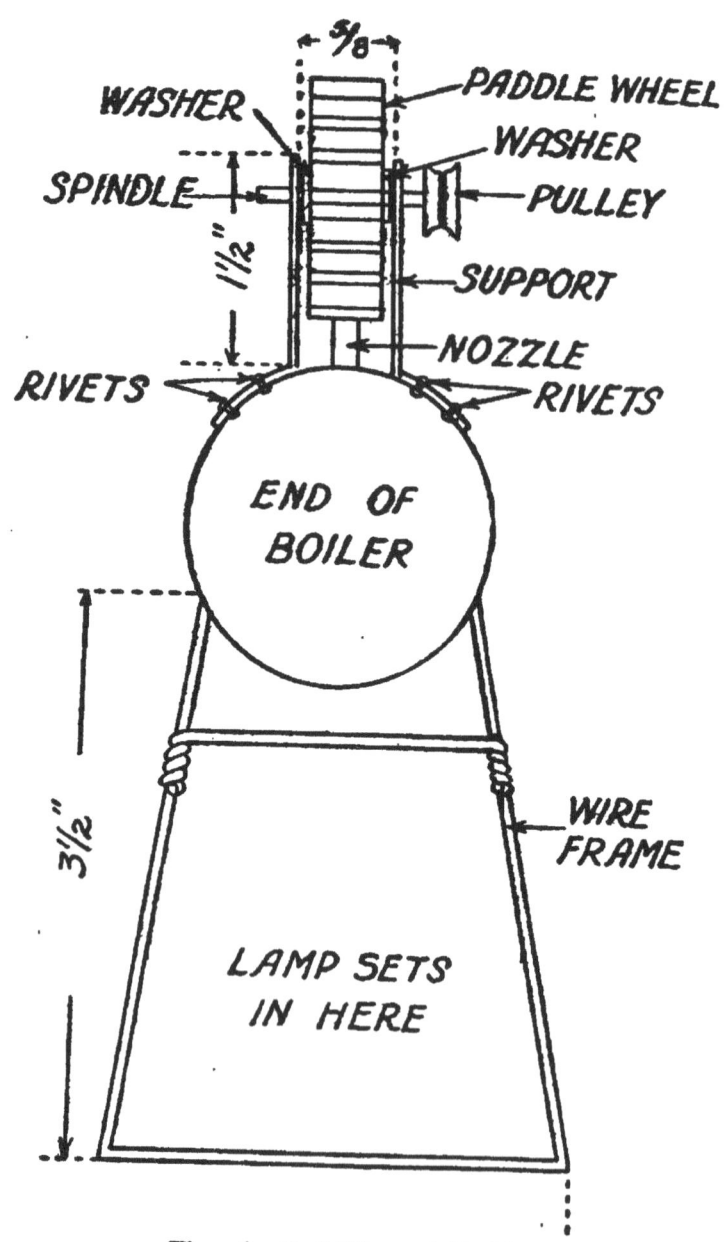

Fig. 16. End View of Turbine

and 1¼ inches long for the spindle, and push it through the hole in the wheel; be sure that it fits tight and sets absolutely true and be careful not to bend the blades.

Slip a washer over each end of the spindle up close to the wheel and then spring the supports apart slightly when you can slip the spindle into the holes in them with the wheel between. Fix a V-shaped pulley to one end of the spindle and the hardest part of your job is done. Make a frame of wire, shaping it as shown in Figs. 15 and 16, and make it high enough to hold the boiler—it is no longer a baking powder can—about three inches away from the surface it sets on.

An Easily Made Alcohol Lamp.—You can make an alcohol lamp without any expense by using an ink bottle; get a squat one, not more than 2 inches high, if you can, though any kind will do. Fit a ⅜-inch tin tube in a hole in the cork and put a cotton wick in it; half-fill the bottle with alcohol, put the cork in it and you have a serviceable lamp.

Running the Turbine.—Half-fill the boiler with clean water and cork up the filler hole tight.

30 *The Boys' Book of Engine-Building*

Set the boiler and wheel on the frame, light the lamp and set it underneath. The boiler will commence to generate steam in five or ten minutes and, when the pressure is great enough, the steam will be projected from the nozzle against

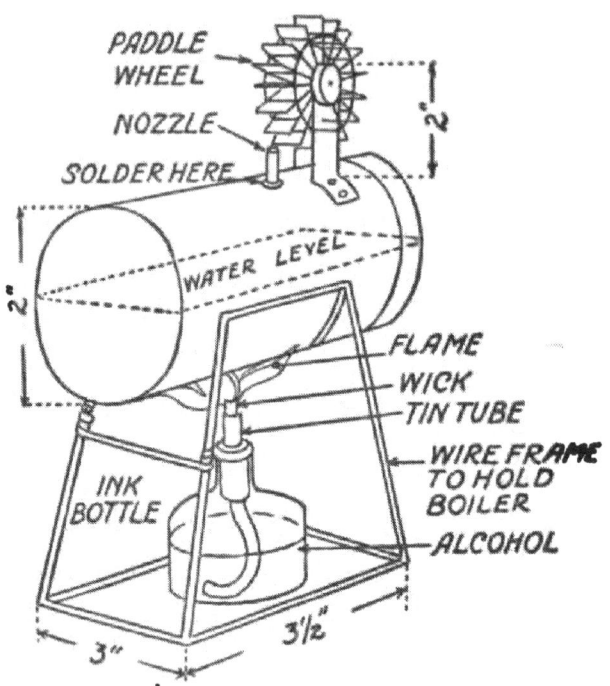

Fig. 17. The Steam Turbine Complete

the blades and the force of it as it strikes them will drive the wheel around at a great rate of speed. The completed turbine is shown in Fig. 17.

A Model De Laval Steam Turbine.—A real

Two Simple Steam Turbine Engines 31

De Laval turbine has curved blades, a special shaped nozzle and a flexible shaft.

The model turbine which I shall describe has curved blades and a regular De Laval nozzle but the flexible shaft need not be made. Since this turbine will run at a high rate of speed the wheel carrying the blades must be very accurately balanced and while you can make such a wheel if

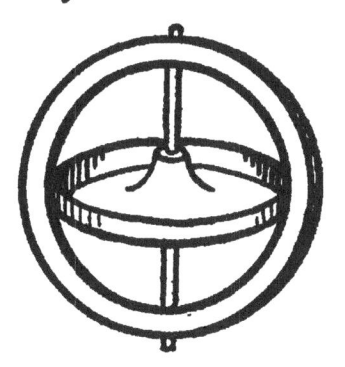

Fig. 18. A Toy Gyroscope

you are a pretty good mechanic, I will tell you an easier way to get one.

You can buy a toy *gyroscope*[1] like the one shown in Fig. 18 for 25 cents. The lead wheel in the *ring bearing* runs very true considering its small cost and you would have hard work to make as perfectly a balanced wheel.

[1] A gyroscope of this kind can be bought of the E. J. Horsman Company, 11 Union Square, New York City.

The Tools You Need.—To make this turbine you need (a) a small machinist's hammer, (b) a pair of tinner's shears, (c) a jeweler's saw frame and saws, (d) a drill stock and drills, (e) a set of taps and dies for cutting screw threads, (f) a jeweller's soldering copper, (g) a couple of files, (h) a pair of compasses for dividing off spaces and scribing circles, and (i) a needle reamer for making tapering holes.

Making the Turbine Wheel.—To make this turbine so that it will be efficient is a harder task than to make any other kind of an engine and my advice is not to start unless you have lots of patience and are willing to work painstakingly.

Begin by taking the wheel shaft of the gyroscope out of its ring bearing and from this moment on be more than careful not to bend the shaft, dent the wheel or to mar either one of them, for if you do the wheel will never run true again.

Take your compasses and draw a circle on a sheet of paper exactly the size of the gyro wheel and divide it into equal divisions ¼ inch apart, or as nearly as you can and still make them come out even as shown at A in Fig. 19. Glue this paper pattern, or *template* as it is called, on the wheel and place one end of the shaft between a couple of pieces of sheet lead and then screw it up in your vise; the purpose of the lead is to keep the rough surfaces of the jaws of the vise from marring the shaft.

Put a saw blade in your jeweler's saw frame and use a blade that is the same thickness as that of the sheet tin, or brass, you intend to use for the blades of the turbine wheel. Saw a slot $3/16$ inch deep in the rim on each line of the paper and be sure to saw each slot straight across the rim of the wheel and straight down on the line of the paper so that each one will be in a *radial line* with the center of the wheel.

Forming the Blades.—Having sawed the slots in the wheel, the next thing is to make the blades. If the wheel of your gyro is $2¾$ inches in diameter, which is the usual size, the circumference of the wheel will be about $8¾$ inches, and since the blades are to be set $¼$ inch apart there will, of course, be 35 blades.

The blades should be made of sheet brass about $1/50$ inch thick and all of them must be as nearly alike as you can make them. Make each blade

⅜ inch wide and cut out the corners to form a shank ⅛ inch wide and 3/16 inch high as shown in the drawing at B, Fig. 19.

File up all the blades nice and smooth and

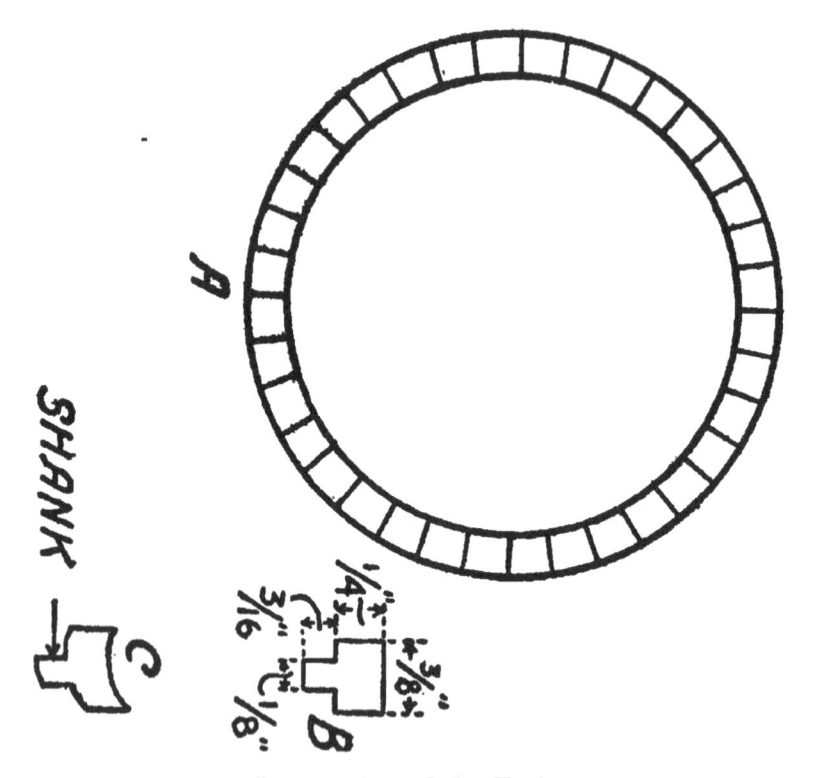

Fig. 19. Construction of the Turbine Wheel

then bend them into the shape shown at C. To do this get a piece of iron rod ¼ inch in diameter and 3 or 4 inches long; file out a half-round groove in a piece of brass or iron until it fits the

¼ inch iron rod exactly. Lay the grooved form on a solid surface, place a flat blade on it, hold the iron rod over it, give the latter a sharp tap with your hammer and the blade will be bent to the proper shape.

Next set the shank of each blade in a slot in the wheel and it should fit in tight. To hold the blades in place a rim must be fitted around them; this can be made by cutting a strip of brass ¼ inch wide and ½ inch longer than the circumference of the wheel with the blades set in it; file down both ends of the strip so that when they are lapped over the joint will have the same thickness as the rest of the ring.

To hold the ring in place on the blades while they are being soldered to it twist a wire around it; then moisten the joint and the tip of each blade where it touches the ring with some soldering fluid. It would be too hard a job to solder each one of the blade tips to the ring with a soldering copper, but you can do it easily and neatly this way: melt one-half a pound of solder in a small shallow pan, dip the rim of the wheel in the solder and turn it slowly around. By so doing the solder will flow evenly and stick to those parts where the soldering fluid has acted.

Fashioning the Nozzle.—The nozzle of a De Laval steam turbine is of special design, as will be explained presently.

To fashion the nozzle get a piece of brass rod ⅜ inch in diameter and 2 inches long. Drill, or have a machinist drill, a 1/16-inch hole through the center of the rod to within ¼ inch of the other end; ream out the hole with a *needle reamer*[1] until the *throat* is 5/32 inch in diameter and the *outlet* is ¼ inch in diameter. This done, drill out the other end of the brass rod for the *intake* of the nozzle with a ¼-inch drill to a depth of ¼ inch, when it will meet the small end of the hole which forms the throat.

Saw off the other end of the nozzle at an angle of 15 degrees, as shown in the cross-section view A in Fig. 20, and the phantom view at B; this will make the *outlet,* or hole, for the steam to be thrown on the blades an *ellipse* which measures ¼ inch wide and ¾ inch long. File the end smooth and have a machinist or a steam-fitter thread one end of it and screw a standard sized

---

[1] A thin, round tool for enlarging and tapering holes.

Two Simple Steam Turbine Engines

union, that is a pipe coupling, on to it so that it can be joined to the supply pipe of the boiler. This completes the nozzle.

The Turbine Wheel Case.—The wheel can be left exposed and the nozzle mounted on a support, but it is better to enclose it so that the ex-

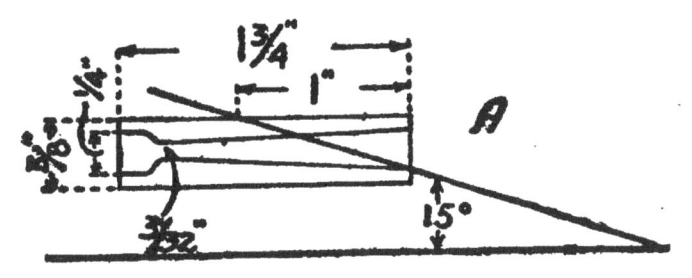

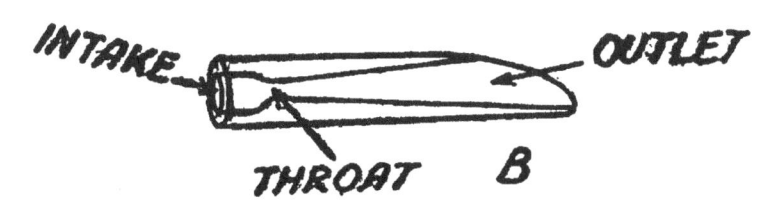

Fig. 20. Construction of the Turbine Nozzle

haust steam can be piped away from it. A metal case or *housing* as it is called, can be made by cutting out two metal plates 1/16 or 1/8 inch thick and 4⅛ inches wide and high and rounding the upper halves as shown at A in the side view, Fig. 21.

Clamp these plates together, or screw them up

in your vise, and drill three ⅛-inch holes, one in each corner and one at the top, as shown by the small dotted circles. Next drill a ⅛-inch hole exactly in the center of both plates for the shaft. In one of the plates drill three ¼-inch holes in a line so that they just meet and have the middle one 1½ inches above the center, or shaft, hole; file these holes out with a small half-round file so that the beveled end of the nozzle will fit snugly into it at the angle shown at B in Fig. 21 and solder it there.

Brace the nozzle with a twisted plate formed of a strip of brass 1/16 inch thick, ½ inch wide and about 1⅜ inches long; drill a ⅜-inch hole in one end; cut a slit on each side of it ⅛ inch deep half way between the ends, and then twist them at right angles to each other as shown at C in Fig. 21. Slip the end with the hole in it over the nozzle and solder the other end to the case plate as shown at A.

Drill a ½-inch hole near the lower side of the other case plate for the exhausts, and this ought to have a pipe soldered to it. Finally, get three 6-32 brass machine screws 1 inch long with a washer and a nut on each one, also get a piece

Two Simple Steam Turbine Engines 39

of brass tubing just large enough to slip over one of these screws and cut off three lengths of the tube each $1\frac{1}{16}$ inch long.

To assemble the case put a screw through each

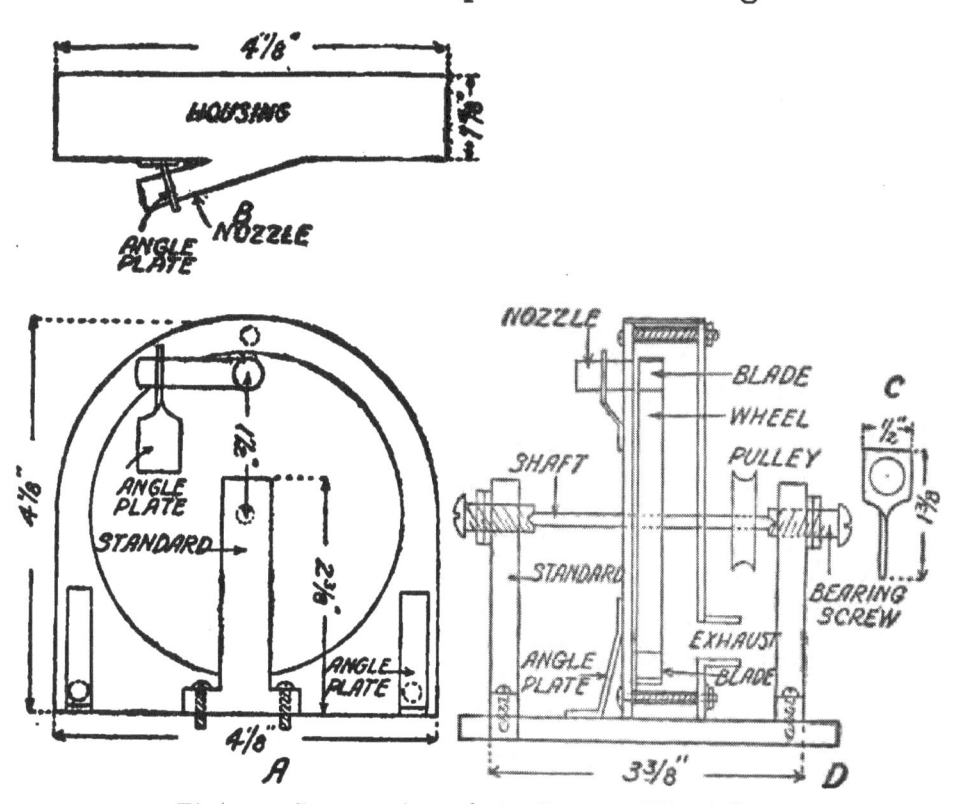

Fig. 21. Construction of the Turbine Wheel-Case

hole in the plate which carries the nozzle; slip one of the tubes over each screw to keep the plates apart and insert the shaft of the wheel in the center hole of the nozzle plate. Next set the case plate with the exhaust hole so that the screws will go through the screw holes in it; put on a washer and screw a nut on each screw.

Cut out four strips of brass for the angle plates $5/16$ inch wide and 1½ inches long, and drill a hole in one end of each strip; bend these as shown in the cross-section view D in Fig. 21 and solder them to the sides of the case plates near the corners of the case as shown at A and D. These are to be screwed to the bed of the machine to hold the case rigidly in place.

The Bearings for the Shaft.—You can buy the bearings for the shaft or you can make them yourself. An easy way to do the latter is to cut a pattern out of soft pine and cast a couple of them in type metal or better, have a molder cast them in brass for you.

Each bearing consists of a *standard,* that is a support, ¼ inch thick, ½ inch wide and 2⅜ inches long with feet on one end as shown at A and D. Smooth up the castings with a file, then drill a $3/16$-inch hole ⅜ inch from the plain end and thread these holes with a tap to take a ¼-inch screw. Get two brass machine screws, each ¾ inch long, and drill out the ends of the screws

Two Simple Steam Turbine Engines 41

$3/32$ inch deep. Also drill a ⅛-inch hole in each foot so that the standard can be screwed to the bed.

Mounting the Wheel.—Get a piece of *boiler plate,* or any kind of a smooth iron or brass plate, about ¼ thick, 4 inches wide and 5 inches long for the bed.

Drill four ⅛-inch holes in this plate for the bearing standards, two on each side; have each pair of holes $15/16$ inch apart and ⅜ inch from the other as shown in A and B. Tap out these holes and screw the standards tight to the bed.

Set the shaft of the wheel between the bearing screws and drill a $3/32$-inch hole through the bed at the places where the angle plates of the wheel case rests on it; thread them with an 8-32 tap and screw the angle plates down to the bed. This completes the steam turbine proper and, when done, it will look like Fig. 22.

The Reduction Gears.—When supplied with enough steam, this turbine wheel will run at about 10,000 revolutions per minute, and this, of course, is much too speedy for any earthly use.

To reduce the speed to something like that of an ordinary engine a *reduction gear* is needed, that is a very small gear, or *pinion* as it is more often called, is made to turn a very much larger gear and, of course, this one turns more slowly, the rate depending on the relative sizes of the two

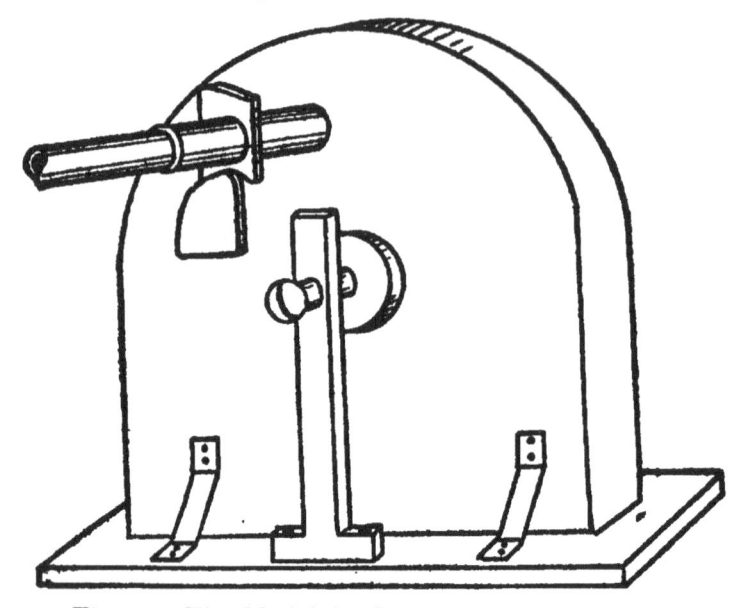

Fig. 22. The Model De Laval Turbine Complete

gears, or their *ratio* as it is called. A *single reduction* gear is shown in Fig. 23.

Sometimes there is a second small gear fixed to the shaft of the large gear; this small gear is made to turn another large gear and this gives you a *double reduction* gear. You can rig up a reduction gear by using the wheels and pinions taken from an old clock or you can buy them

Two Simple Steam Turbine Engines 43

from dealers in model maker's supplies. For a boiler large enough to run this turbine see Chapter V.

How the Steam Turbine Works.—You may not have noticed it, but the De Laval turbine

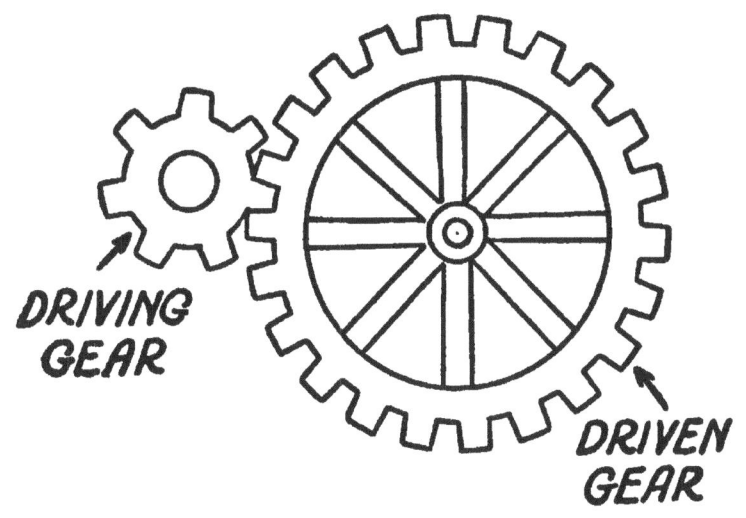

Fig. 23. A Pair of Single Reduction Gears

is both an impulse turbine like Branca's and it is also a reaction turbine like Hero's.

The Action of the Blades.—If you will turn back to Fig. 11 you will see that the steam from the nozzle, when it strikes the blades of the wheel, pushes it forward by the impulsive force of the energy that is stored up in it. You will also observe that, after the steam strikes the *curved blades,* it is thrown out *against the air* and this causes the force of the steam to react on the wheel just as it does in Hero's engine.

How the Nozzle Works.—When steam is allowed to escape through a simple hole, or *orifice,* a considerable amount of the energy in it is wasted in setting up little whirls and other commotions in it and this leaves only a small portion of its useful energy to drive the wheel with.

To prevent these untoward actions as well as to make the steam *expand*[1] in the nozzle, that is to change the energy which exists in it as latent heat into the energy of motion, a special shaped nozzle must be used and this is the purpose of the one De Laval designed.

When the steam from the boiler reaches the throat of the nozzle, it falls to about half of its *initial pressure,* that is the pressure delivered by the boiler to the nozzle. After passing through the throat the steam begins to expand and thus, while it decreases in density, in pressure and in temperature, the particles of which it is formed increase in speed and hence in power.

[1] See Chapter IX, "The Stuff that Steam is Made Of."

CHAPTER III

A SIMPLE PISTON STEAM ENGINE

The Oscillating Cylinder Engine—The Tools You Need—The Parts of the Engine: The Cylinder; The Piston; The Crank, Crankshaft and Flywheel; The Engine Bed—Assembling the Engine: Oiling the Engine; Testing the Engine—How to Make the Boiler—How to Run the Engine: About the Lamp—How the Power Plant Works—Where to Buy Materials.

Before you build any of the larger and more complex engines which follow, you should by all means make the parts of and assemble the simple single-acting steam engine described in this chapter.

By building this one you will get a clear idea of the way steam acts and how an engine works and this will be of great help to you when you come to the more powerful *slide-valve* engines.

The Oscillating Cylinder Engine.—An oscillating cylinder engine is one in which the cylinder is pivoted in the middle and by rocking up and down like a druggist's balance, or *oscillating* as it is called, allows the steam from the boiler to pass into and escape from a single *port*, that is a hole, or opening, in one end of the cylinder.

This kind of an engine is *single acting*—that is the steam is admitted into one end of the cylinder and hence there is only one power stroke to each revolution of the crankshaft. It can be made *double acting*—that is the steam can be admitted into each end of the cylinder alternately and so gives two effective power strokes to every revolution of the crankshaft.

The oscillating cylinder engine which I shall tell you how to build is a *single-acting* one and I have chosen it because it is the easiest kind to make.

The Tools You Need.—You will need very few tools to make this little engine but though a couple of them are rather expensive you ought to have them anyway as they will prove useful for a thousand and one other things.

Get (1) a small *machinist's hammer;* (2) a *hand drill stock* and half a dozen *twist drills;* (3) a couple of *files;* (4) a pair of *tinner's shears;* (5) a small *bench vise;* (6) an *alcohol blow torch;* (7) a *jeweller's soldering copper;*

A Simple Piston Steam Engine

(8) ten cents' worth of *wire solder,* and (9) make some *soldering fluid* by dissolving *zinc clippings* in a little *muriatic acid.*

The Parts of the Engine.—There are six principal parts to this engine: (1) the *cylinder;* (2) the *piston;* (3) the *piston rod;* (4) the *crankshaft;* (5) the *flywheel,* and (6) the *bed.*

The Cylinder.—To make the cylinder get a piece of brass or copper tubing [1]—it must be smooth inside—and have it ⅜ inch in diameter, inside measurement, 1⅜ inches long and with as thick a *wall* as you can get it.

File off the cylinder on one side so that it is perfectly flat and then drill a hole through the middle of the flat side 1/16 inch in diameter for the pivot and another hole of the same size ⅛ inch from the end, all of which is shown at A in Fig. 24.

Take a 6-32 screw, cut off the head and file the end of it down until it fits in the middle hole; moisten it with a little soldering fluid, turn the flame of your blow torch on it for a moment and then touch the end of the wire solder to it when the latter will run around and join the two pieces

---

[1] Seamless tubing is the best.

of metal firmly together. Make a spiral spring of brass wire just large enough to go over the pivot and screw a nut on the latter.

To make the *heads* of the cylinder cut out two disks of sheet brass, or you can use sheet lead which is easier to work, and drill a ⅛-inch hole

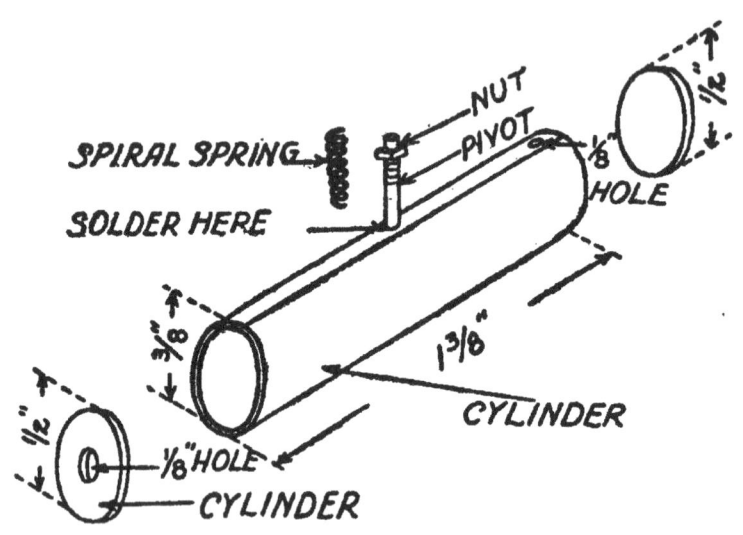

Fig. 24 A. The Cylinder and Cylinder Heads

in the center of one of them. These heads are to be soldered to the ends of the cylinder, and this can be done better after the piston is made.

The Piston.—Because the piston must slide to and fro in the cylinder very smoothly and yet close enough to prevent the steam from leaking

A Simple Piston Steam Engine 49

past it, the making of the piston is a most particular job.

Since the inside of the cylinder is ⅝ inch in diameter, the *piston* must be nearly the same size. Take a piece of brass about ³⁄₃₂ inch thick and *scribe,* that is scratch a circle on it with a pair of compasses ⅝ inch in diameter; drill a ⅛-inch hole in the center and file down the piece of brass

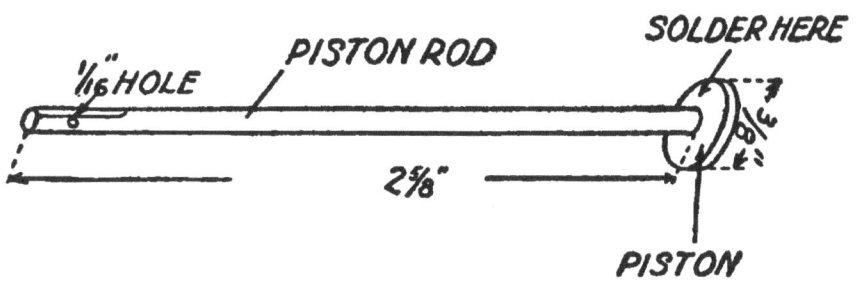

Fig. 24 B. The Piston and Piston Rod

until it is a perfect circle and fits the *bore* of the cylinder accurately.

You can make the *piston rod* of a straight piece of iron wire ⅛ inch thick and 2⅝ inches long; file off the sides of one end a little and then drill a ¹⁄₁₆-inch hole through it ³⁄₁₆ inch from the end as shown at B in Fig. 24. This done, fit the other end of the piston rod into the piston and be sure to get it in straight, and then solder them together.

50 The Boys' Book of Engine-Building

Having the cylinder and the piston, the next thing to do is to slip the cylinder head with the hole in it over the piston and put the latter in the cylinder with the cylinder head on the end of the cylinder opposite the port. Now solder

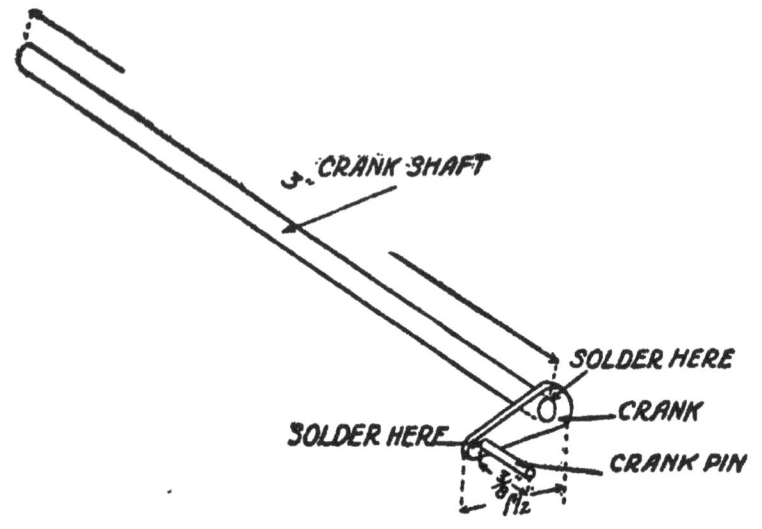

Fig. 25 A. The Crank and Crankshaft

the head to the cylinder using your soldering copper to do the job.

Be careful that the head is on so that the piston will work forth and back without binding. When you have done this, go ahead and solder on the other head to the cylinder and the hardest part of your engine is made.

The Crankshaft, Flywheel and Pulley.—A

A Simple Piston Steam Engine 51

straight piece of wire $3/32$, or $1/8$ inch thick and 3 inches long will serve for the crankshaft, that is that part of the engine which revolves, to which the piston rod is connected and on which the flywheel and pulley are fixed.

For the *crank,* see A, Fig. 25, cut out a piece

Fig. 25 B. A Spoked Fly-Wheel

of brass $1/16$ inch thick and $1/2$ inch long, and make one end $1/4$ inch wide and the other end $1/8$ inch wide; drill a $1/16$-inch hole in the small end and a $1/8$-inch hole in the large end; fit in and solder one end of the crankshaft to the crank and solder a $1/16$-inch wire, $3/8$ inch long, into the other end for the *crank pin.*

Almost any kind of a wheel will do for the flywheel, see B, but it should be pretty large, say 3 inches in diameter, and it should be rather heavy, say about 2 ounces, and the larger it is for its weight the smoother the engine will run.

You can cast a flywheel with a little *pulley* on it by sawing out a wooden pattern first, making a mould of *plaster paris* and pouring in some melted *type metal*.[1] An easier way is to buy one already cast.

The Engine Bed.—This is a block of wood 1 inch thick, 2½ inches wide and 5 inches long on which to mount the engine; sandpaper it smooth and give it a coat of red paint.

Cut out a *frame* of a piece of thick sheet brass, or heavy tin will do, 4 inches square to the shape shown in Fig. 26. Drill a ³⁄₃₂-inch hole in the center of the circles marked *inlet port* and *exhaust port* and in each of those marked *bearing;* also drill a ⅛-inch hole in the circle marked *cylinder pivot* and in the *screw hole* circles.

Now bend up the sides of the tin or brass on the dotted lines to form the frame and solder a piece of brass pipe ⅛ inch in diameter and 6

[1] You can use lead, but it shrinks on cooling.

A Simple Piston Steam Engine 53

inches long over the inlet port on the *inside* of the *valve plate*.

Assembling the Engine.—You have now reached the last and most interesting stage of

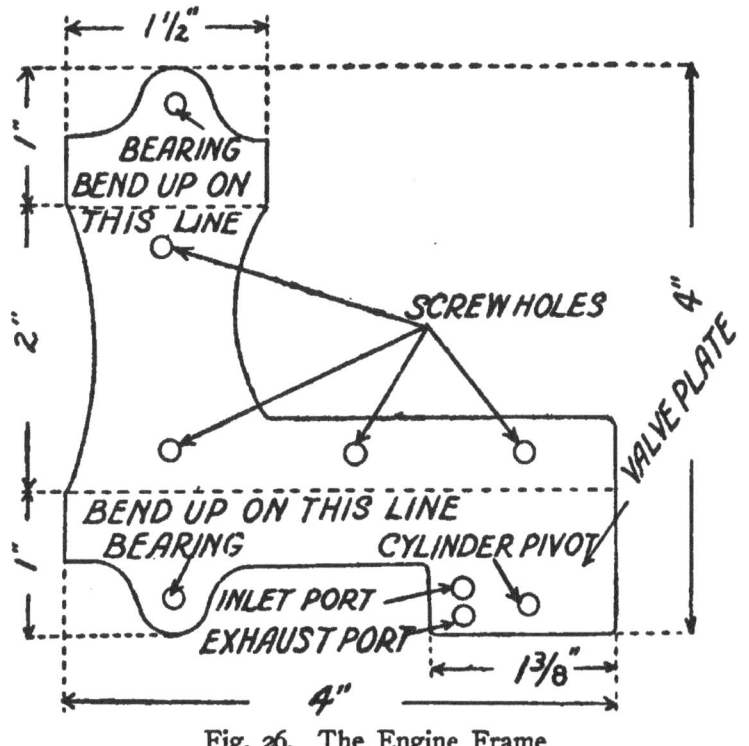

Fig. 26. The Engine Frame

your work, the *assembling* of the parts. Begin by screwing the frame to the bed of wood and have the valve plate and bearing on the side with it *flush*, that is even, with one of the long edges of the bed.

54 The Boys' Book of Engine-Building

Next slip the crankshaft through the bearings of the frame and then fit the flywheel and pulley to the other end and, if these are made of metal, solder them on to the end of the shaft. Slip the crank pin through the hole in the end of the piston rod; put the cylinder pivot through the pivot

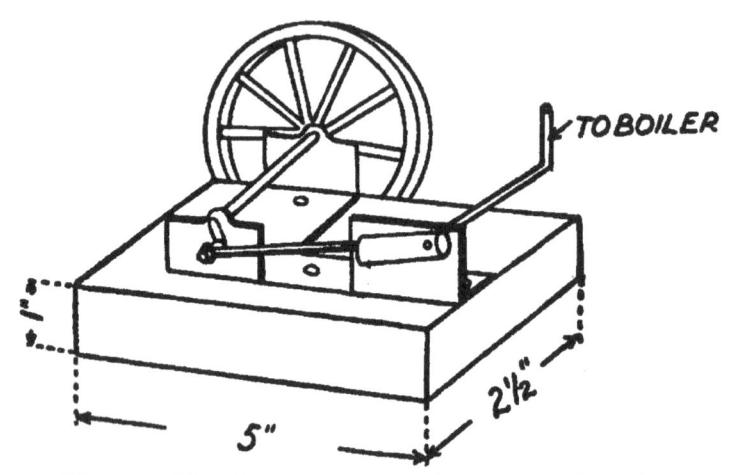

Fig. 27. The Oscillating Cylinder Engine Complete

hole in the frame, set the spiral spring over the pivot and screw on the nut. This arrangement permits the cylinder to oscillate freely and at the same time holds it close to the valve plate. The engine complete and ready to run is shown in Fig. 27.

Oiling the Engine.—To keep the steam from leaking between the port of the cylinder and the

A Simple Piston Steam Engine

inlet port of the valve plate, *lubricate* it, that is oil it, with a drop of sewing machine oil and also put a couple of drops of oil into the cylinder as this will make it run with less friction and keeps the steam from getting past it; finally, put a wee drop of oil on each of the bearings and your engine is ready to run.

Testing the Engine.—Having put your engine together, the next thing to do is to try it out to see if it is in working order. All you need to do to find out if it will run is to give the flywheel a start and blow through the inlet pipe. If you have made it right, it will run as long as you blow. When it runs to your satisfaction, your next need is a boiler.

How to Make the Boiler.—To make a boiler that will raise enough steam to operate this little engine is a simple matter.

Make a can 2¼ inches in diameter and 6 inches long of heavy sheet tin. To do this cut out a piece of tin 6 inches wide and 7 inches long; bend it on a broomstick until it is a cylinder and solder the edges together thus, making a *lap seam*.

Cut out two disks of tin 2¼ inches in diameter

and with four ears on each one as shown at A in Fig. 28. Bend the ears over and solder the disks to the ends of the boiler and do it well. Now drill a ⅛-inch hole through the seam 2½ inches from one end and a ¼-inch hole also through the seam 1½ inches from one end.

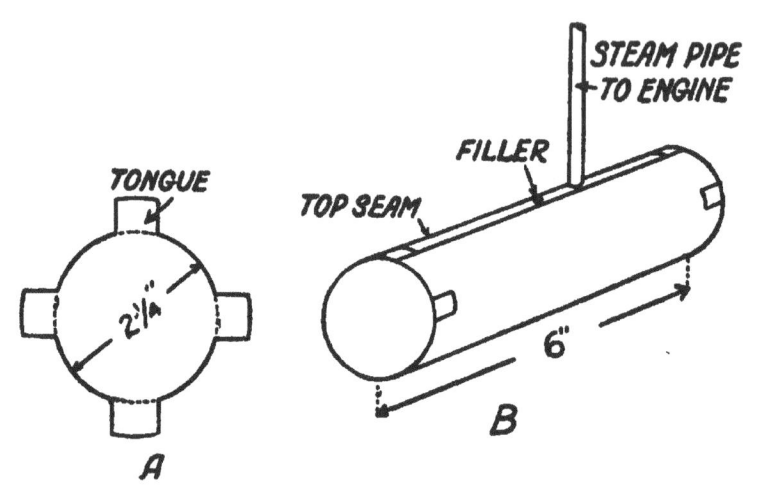

Fig. 28. Construction of the Boiler

Make a box of heavy tin, or, better, of sheet Russian iron, for the boiler to rest on, 3 inches high, 3 inches wide and 5 inches long. This can be made of a single piece of sheet metal by marking out a *template,* that is a pattern as shown at C in Fig. 28. Draw the pattern out full size on a sheet of paper and paste it on the tin or sheet iron. Drill a dozen ⅛-inch holes

A Simple Piston Steam Engine   57

near the top on the sides and a dozen or more holes of the same size near the bottom in the middle so that the flame can get plenty of air and the heat will be kept in at the same time.

Bend the iron or tin into the form of a box and then bend out the edges at the bottom and bend in the edges at the top.   Now make a wood

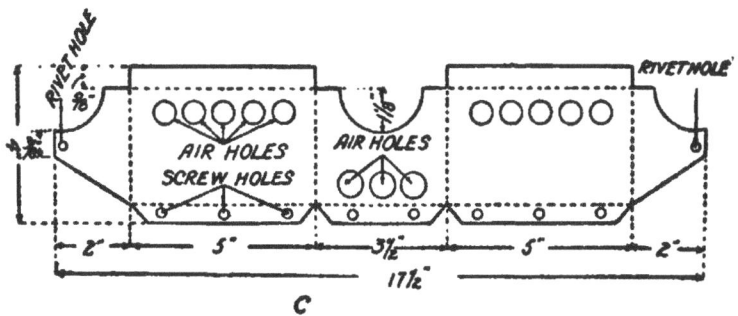

Fig. 28 C. The Fire-Box

base 1 inch thick, 5 inches wide and 7½ inches long. Screw the flanges of the box on the base leaving a margin of ½ inch all round. Set the boiler in the concave ends of the box, as shown in Fig. 29, and bring the steam pipe up from the engine and bend it so that it will fit close to the small hole in the top of the boiler and solder it fast.

Half-fill the boiler, through the large filler hole, with clean water and put a cork in the hole; the

58 *The Boys' Book of Engine-Building*

cork will not only keep the steam in, but it will act as a *safety valve,* for it will blow out before the boiler blows up should the steam pressure get too high.

There is only one more part to make before your miniature power plant is complete and this

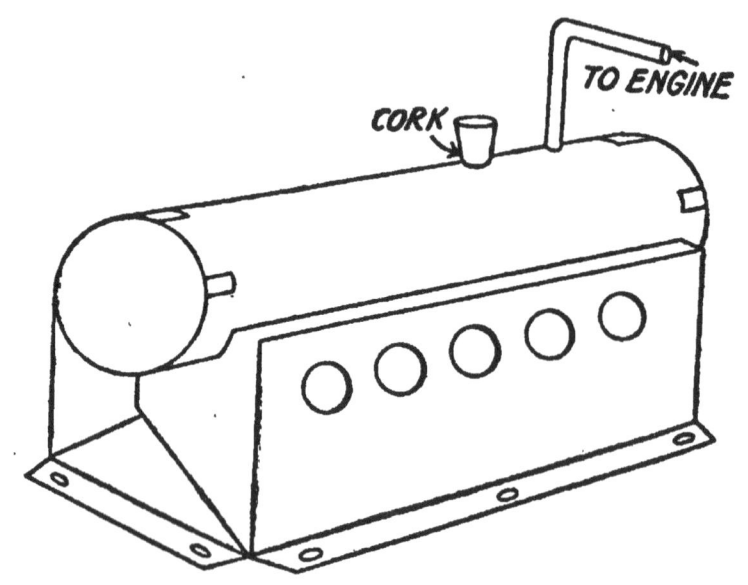

Fig. 29. The Boiler Mounted on the Fire-Box

is the *furnace* to heat the water in the boiler. Get, if you can, or make, if you have to, a tin salve box about ¾ inch high and 2 inches in diameter. Drill, or punch, a ⅜-inch hole through the center of the lid and solder a tin tube ⅜ inch in diameter and ⅝ inch high in it for the burner;

drill a 1/16-inch hole in the outer edge of the lid for a vent; make or buy a braided cork wick to fit the burner, about 1¾ inches long, and put it through the tube.

How to Run the Engine—*Filling the Boiler.*—By filling the boiler with boiling hot water it will not only make steam quicker but it will save alcohol. Be sure the water is clean or else the steam port of the engine may get choked up and then the engine will run slowly or stop altogether. And *be careful* the boiler does not run dry.

About the Lamp.—Have the wick just a trifle above the tube and half-fill the lamp with wood alcohol. Wipe off any alcohol that may be spilled on it in filling and then light the wick.

If the water is boiling hot when you put it in the boiler and the flame of the lamp is burning blue, it will take about five minutes to get up steam. At the end of this time turn the flywheel around several times to clear out the steam which has condensed in the cylinder and, if the steam is under sufficient pressure, your engine will run like sixty.

How the Power Plant Works.—First the heat of the lamp changes the water in the boiler into steam and, when the latter is hot enough, it is under pressure.

Now when the piston is at the back end of the cylinder, the port is then just even with the opening of the pipe in the valve plate; the steam under pressure rushes into the cylinder and forces the piston forward; as the crankshaft turns round the rear end of the cylinder is raised until the port is even with the exhaust hole in the valve plate and this permits the used steam to escape into the air.

By this time the crankshaft will have made one complete revolution and the *cycle of operation* will begin all over again. It is indeed a pretty mechanical movement.

Where to Buy Materials.—You can buy a cylinder and piston for thirty cents and a flywheel for the same price ready-made of the Weeden Manufacturing Company, New Bedford. Mass., or you can buy an engine and a boiler complete, very like the one I have described, of the above firm for $1.50.

CHAPTER IV

A 1/24-H. P. HORIZONTAL STEAM ENGINE

The Tools You Need: The Drawing Tools; The Wood Working Tools; The Metal Working Tools—Drawing the Plans of the Engine—The Parts of the Engine—How to Make the Patterns: The Cylinder: The Cylinder Heads, The Piston; The Steam Chest, The Slide Valve; The Cross Head Guide; The Crank; The Eccentric; The Pillow Blocks; The Bed Plate—Moulding the Parts in Metal—Finishing the Castings: The Cylinder; The Steam Chest; The Piston and Piston Rod; The Slide Valve and Valve Stem; The Cross Head Guide: The Cross Head Guide Block, The Rocker Arm, The Valve Stem Bearing; The Connecting Rod; The Crank and Crank Shaft; The Eccentric Rod; The Eccentric; The Flywheel; The Pillow Blocks—Mounting the Engine on the Bed Plate—Setting the Engine on Its Bed—The Auxiliary Parts—How the Engine Works—Calculating the Horse Power of a Steam Engine.

There are three ways to go about making an engine of this size and kind and you can choose the one that suits your pocket and your purpose the best.

The easiest way is to buy a complete set of castings (See Appendix D), have such machine work done on them as may be needed and then assemble the parts.

The second way—and this is my idea of building the engine—is to make patterns of all the parts yourself, have them cast in iron or brass at a foundry, get a machinist to bore the cylinder and the ports in it, and bore and turn the flywheel, when you can do the rest of the work and assemble it without trouble.

A cheaper but less satisfactory way is to make the patterns as before and cast them yourself in type-metal. Castings of this metal are very sharp and yet quite soft, hence they can be easily worked. When it is done in this way, the engine will make a good working model, but you will have to handle it very gingerly to keep it from getting marred.

The Tools You Need.—These consist of (1) a few drawing tools; (2) some woodworking tools, and (3) a fair assortment of metal working tools.

The Drawing Tools.—You can get all the drawing tools you need for a dollar or more. The necessary ones are (a) an 11 x 15½ inch drawing board; (b) a good accurate rule—an architect's triangular boxwood scale is the best; (c) a 30° triangle; (d) a T square as long as the board; (e) a pair of 4-inch dividers; (f) a pair of 4-inch compasses; (g) an eraser, and (h) a good, medium-soft lead pencil.

The Wood Working Tools.—These can be limited to (a) a small block plane; (b) a scroll saw frame and saw blades; (c) a small back saw; (d) a miter box; (e) a brace and ¼-, ½ and ⅞-inch bits; (f) a try square and (g) a good pocket-knife.

The Metal Working Tools.—The following list of tools includes about everything you need and while you might get along with a few less you could, of course, use several more.

Get (a) a jeweller's saw frame and saws; (b) a jeweller's hammer; (c) a small machinist's hammer; (d) a center punch; (e) a pair of spring dividers; (f) a pair of tinner's shears; (g) a hand drill stock and an assortment of twist drills; (h) an alcohol lamp; (i) a jeweller's soldering copper; (j) a pair of flat-nose side cutting pliers; (k) a couple of screw drivers; (l) some files; (m) a set of taps and dies; (n) a carborundum oil stone; (o) an oil can filled with sewing machine oil, and (p) a vise.

Drawing the Plans for the Engine.—The first thing to do is to draw all the separate parts of the engine shown in Fig. 31 on a sheet of paper full size and mark on the dimensions. You will then

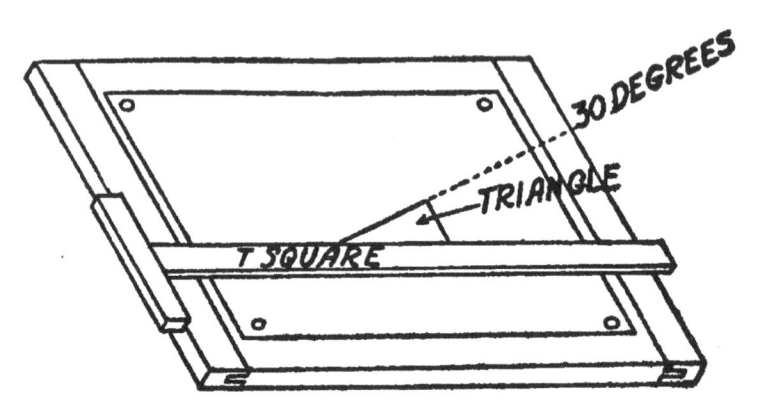

Fig. 30. T-Square and 30 Degree Triangle on Drawing Board

have a set of drawings that you can work from much better than those shown in the book.

I have made the drawings of the parts in what is called *isometric perspective* so that you can clearly see the design, construction and dimensions at a glance. To make the drawings fasten a sheet of paper on the board with *thumb-tacks;* lay your T square on the board with the 30 degree triangle on the upper edge of it as shown in Fig. 30; you can then draw 30 degree lines which are used for *isometric* pictures.

The Parts of the Engine.—This little horizontal engine is formed of the following parts: (1) the *cylinder;* (2) the *steam chest;* (3) the *piston* and *piston rod;* (4) the *slide valve* and *slide valve stem;* (5) the *cross head guide;* (6) the *cross head guide block;* (7) the *eccentric rod rocker arm;* (8) the *connecting rod;* (9) the *eccentric rod;* (10) the *crankshaft* and *crank;* (11) the *eccentric;* (12) the *flywheel;* (13) the *pillow blocks;* (14) the *bed plate,* and (15) the *bed,* all of which will be described in detail as we go along.

How to Make the Patterns.—All the patterns are made of wood which must be well seasoned; soft pine without knots and having a straight grain is the easiest wood to work. You can build up each part of as many pieces of wood as you want, but be sure that they fit and glue them well together.

Each part must be built up exactly as you want the casting to appear when finished, only a trifle larger to allow for shrinkage and machining; and to make the pattern *draw* from the mould smoothly sandpaper it with fine sandpaper until it seems to be one solid piece; but in sandpapering it be careful not to round off any edges that should be sharp. After sandpapering give the pattern a couple of coats of *shellac varnish* and it is then ready for the moulder.

The Cylinder.—Take a piece of clear pine 1⅜ inches square and 2 inches long and bore a hole through the middle of it ⅞ inch in diameter; plane it down until it is 1¼ inches across on top, ⅝ inch wide and flat on the sides and 1¼ inches in diameter on its rounded side as shown at A in Fig. 31.

This will make the *wall* of the cylinder, that is its thickness, 3/16 inch thick. You won't need to drill the *port holes* in the pattern as it can be done to better advantage in the casting when it is made.

Saw out two rings of wood ⅛ inch thick and make the inside diameter of each one ⅞ inch and the outside diameter 1¾ inches; glue a ring to each end of the cylinder for the flanges and screw them up in a clamp or a vise until the glue has set.

Next saw out two feet, as shown at B, ¼ inch

Fig. 31. The Parts of an Engine

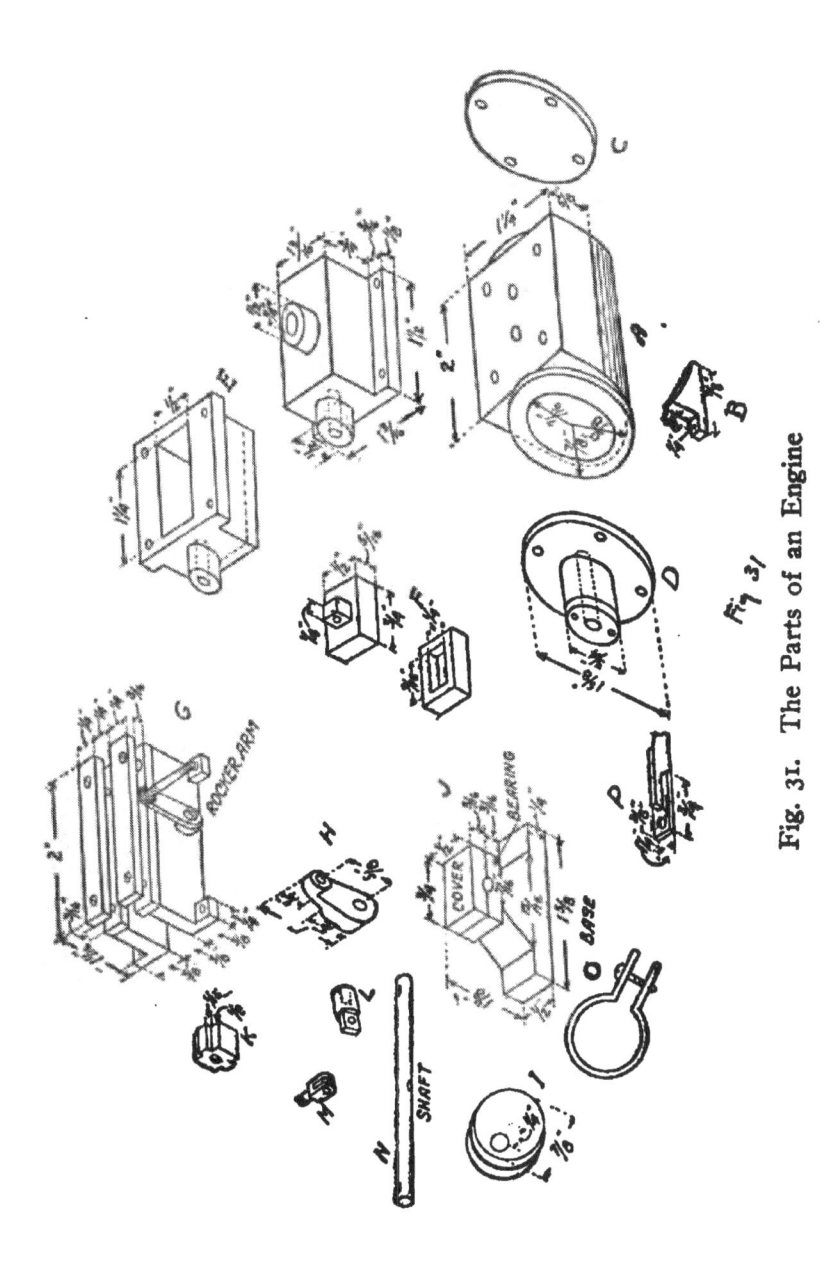

wide, ⅝ inch high and 1⅞ inch long; make the foot proper ⅛ inch thick, round the top of the pattern so that it will fit the cylinder and glue a foot to each side of the latter as shown in Fig. 32. When the glue has set, sandpaper and shellac the pattern and it is ready for the moulder.

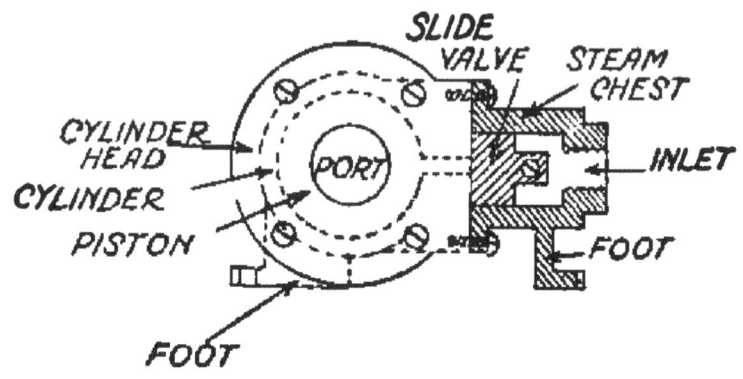

Fig. 32. An End View of the Engine

The Cylinder Heads.—It is much easier to make patterns for the cylinder heads than it is to cut them out of sheet metal. The back cylinder head, see C, is simply a disk ⅛ inch thick and 1¾ inches in diameter.

The front cylinder head is the same size but it has a ring on it 7/16 inch long with a bore ½ inch in diameter and an outside diameter of ¾ inch as shown at D. This ring forms the *stuffing box* and allows the piston to be packed, which keeps the steam from escaping around it. The small ring that holds the packing in can be cut out of a piece of sheet brass.

The Piston.—Saw out a disk of wood ¼ inch thick and 1⁵⁄₁₆ inches in diameter for a pattern for the piston.

The Steam Chest.—This is simply a box open at the bottom. It is ⅛ inch thick, ¾ inch high, 1³⁄₁₆ inch wide and 1½ inches long, outside measurement, as shown at E.

Glue a strip ⅛ inch thick, ³⁄₁₆ inch wide and 1½ inches long on each side of the box along the bottom; this will make it just the width of the top of the cylinder but ½ an inch shorter in length. Now glue a disk ⅛ inch thick and ½ inch in diameter in the center and on top of the steam chest and then glue a ring ¼ inch long with a ⁵⁄₁₆ inch bore and an outside diameter of ⁹⁄₁₆ inch on one end of the chest. This is for the stuffing box.

The Slide Valve.—Saw out a block ⁵⁄₁₆ inch thick, ½ inch wide and ¾ inch long; cut out one side with a chisel, or your knife; so that it will be ³⁄₁₆ inch deep, ¼ inch wide and ⁹⁄₁₆ inch long. Cut out a block ¼ inch square and glue it to the center of the top of the slide valve as shown at F; to this is fastened the slide valve rod.

The Cross-Head Guide.—You only need one pattern for the supports, see G, and have two castings made of it. Saw out a strip of wood 3/16 inch thick, 5/8 inch wide and 2 inches long and glue a block 1/8 inch thick, 1/4 inch wide and 3/8 inch long to one edge at each end for feet so that the castings can be screwed down to the bed-plate.

The Crank.—From a piece of wood 1/2 inch thick and 1 inch long cut out a crank as shown at H. Make one end 1/2 inch wide, the other end 1/4 inch wide and round off both ends. The finished metal crank is used to connect the connecting rod to the crankshaft.

The Eccentric.—For this pattern cut out a disk 3/8 inch thick and 7/8 inch in diameter as shown at I.

The Flywheel.—You can buy an iron flywheel in the rough with a 7/8-inch face and 5 inches in diameter for 25 cents.

If you want to make a pattern and have it cast, so much the better, but you must expect to

A 1/24-H. P. Horizontal Steam Engine 71

spend a lot of time on it. Make it with six spokes and saw it out with your scroll saw.

The Pillow Blocks.—These are to support the crankshaft and to provide the bearings for it. Each block has a cover as shown at J.

To make a pattern saw out a block of wood ½ inch wide, 1 inch high and 1⅝ inches long to the

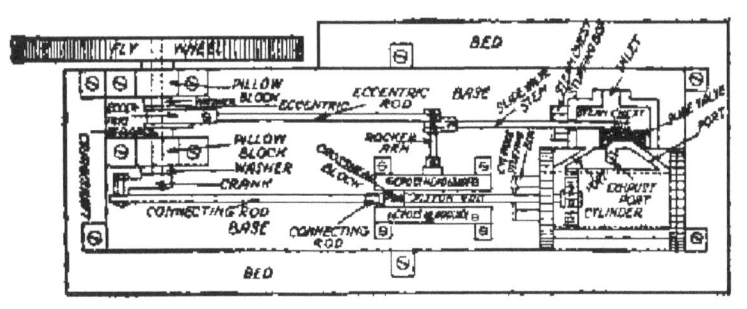

Fig. 33 A. Top View of the Engine with Cylinder and Steam Chest in Cross Section

shape shown at J. Have the base ¼ inch thick and the bearing 3/16 inch thick, ½ inch wide and ¾ inch long and make the cover the same size and cast them separately.

The Bed Plate.—This is the plate to which the cylinder and other parts of the engine are bolted and it should be cast in iron.

Make a frame of ⅛-inch stuff, 1 inch deep, 3 inches wide and 9⅞ inches long and put a top on it ⅛ inch thick. Cut out six blocks ¼ inch thick and ⅜ inch square and glue one to each corner of the bed plate at the ends and one to each side in the middle. These are to permit the bed plate to be bolted to the bed.

Moulding the Parts in Metal.—When all of the patterns are finished, take them to a foundry and have them cast in iron, or, better, in brass.

As the amount of metal in the castings is small the cost will be about the same for either. On the whole brass castings are the best because they are finer grained, easier to work and make a handsomer engine.

Finishing the Castings.—When you get the castings from the foundry they will be a little rough all over and the working parts and those that fit together must be smoothed and trued up either with a file or in a lathe. If you own a lathe fitted with a *slide rest,* you can do all the work yourself, but if you haven't one, then get some genial machinist to help you out.

The Cylinder.—This must be bored out with a 1-inch drill to make it smooth and of the right size. If you have a machinist do this for you, you might as well let him drill the ports and screw holes through the flanges and cylinder heads as well as the steam chest and the steam chest plate.

If you drill the ports yourself, use a ³⁄₁₆-inch drill and start it ⅜ inch from the center of the cylinder on the plate the steam chest sets on; drill the hole at an angle of nearly 40 degrees, until it goes through and meets the bore at a point just inside the flange as shown in the top view, Fig. 33. Then drill another hole of the same size for the other intake port and at the same angle on the opposite side of the plate.

For the exhaust port drill a ⅛-inch hole straight down through the center of the steam chest plate to a depth of ¼ inch; now drill the same sized hole on the other side of the cylinder from that on which the foot is cast and through the cylinder until it meets the first hole. This makes a passage clear through from the steam chest to the open air for the exhaust steam to escape.

Drill four holes in the steam chest plate at the points shown in the side view in Fig. 34 and thread them with an 8–32 tap so that the steam chest can be screwed to the cylinder. Drill a

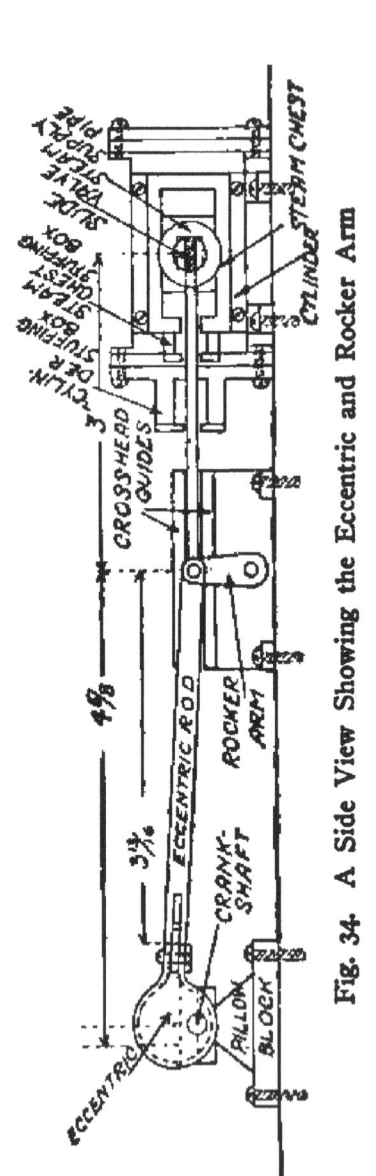

Fig. 34 A Side View Showing the Eccentric and Rocker Arm

3/16-inch hole through the center of the cylinder head with the stuffing box on it.

Next make a ring of 1/16-inch thick sheet brass 3/4 inch in diameter and with a 3/16-inch hole in its center. Drill two 3/32-inch holes through the ring on opposite side of it, drill two 1/16-inch holes in the face of the stuffing box and thread these holes so that the ring can be screwed to it, all of which is shown at D in Fig. 31.

Drill four 3/32-inch holes at equal distances apart through both of the cylinder heads and into the ends of the cylinder; then thread the latter holes so that the heads can be screwed to them.

The Steam Chest.—Drill a 3/16-inch hole through the top of the steam chest for the inlet pipe and thread it to take a 1/4-inch pipe. Drill a 1/8-inch hole through the end with the stuffing box on it for the slide valve stem and drill two 1/16-inch holes in the stuffing box and thread them.

Make a ring 9/16 inch in diameter, cut a hole in it 3/32 inch in diameter and drill two holes near the outer edge so that it can be screwed to the stuffing box. Drill out two holes on one side of the steam chest, and thread them so that the chest can be screwed down to the bed plate.

File the inside edges of the chest true and smooth and finish up the outside of it with a file. Two views of the steam chest are shown at E in Fig. 31.

The Piston and Piston Rod.—The piston casting can be filed down or it can be put in a lathe and then turned down so that it fits the cylinder closely and yet will slip through it without binding; a half round groove should be filed or turned in its face. Drill a 1/8-inch hole through the center of the piston for the piston rod.

A good straight piece of soft steel rod, 1/8 inch in diameter and 3 1/4 inches long, will make a good piston rod; thread both ends of it, screw a nut on one end, slip the piston on and screw another nut on tight. Wrap some soft cotton cord on the piston until the groove is full and soak it in sewing machine oil; work it back and forth in the cylinder until it slides easily.

Now slip the piston rod through the cylinder head with the stuffing box on it and screw the latter to the flange of the cylinder; likewise screw the plain cylinder head on to the other end of the cylinder.

Soak some cotton cord in oil and pack it around

the piston rod in the stuffing box until it is full, and then screw the ring to the latter to hold it in.

The Slide Valve and Valve Stem.—File the sides of the slide valve until it fits the inside of the steam chest to a nicety and file the bottom of it until it slides on the cylinder plate accurately.

Drill a $\frac{1}{16}$-inch hole in the boss of the valve, $\frac{1}{16}$ inch from its lower edge, and thread it to take the valve stem. Make the stem of a piece of soft steel rod $\frac{3}{32}$ inch in diameter and $3\frac{1}{8}$ inches long and thread both ends of it. Put the slide valve in the steam chest, slip the stem through the hole in the chest and screw it into the slide valve; now pack the stuffing box with oiled cord as described above and screw on the ring.

Finally, screw the steam chest to the slide valve plate of the cylinder and the hardest part of your engine is done.

The Cross-Head Guide.—File the top and bottom faces of the guide supports true and smooth and drill two $\frac{1}{16}$-inch holes in the top face of each one $\frac{1}{4}$ inch from the ends and thread them. Also drill a $\frac{1}{8}$-inch hole in each foot and a $\frac{1}{8}$-inch hole in the middle of one of the supports for the *rocker arm*.

Cut off four strips of brass $\frac{1}{8}$ inch thick, $\frac{1}{4}$ inch wide and 2 inches long for the guides. Drill two $\frac{3}{32}$-inch holes through each strip $\frac{3}{32}$ inch from one edge so that when it is screwed to the support it will lap over $\frac{1}{16}$ inch. This done, saw off four pieces of $\frac{3}{32}$-inch brass tubing, $\frac{3}{16}$ inch long, to keep the guides apart.

Now put a 4–36 screw through each hole in one of the guide strips, slip a tube over each screw, slip another guide strip over the screws and screw them to the support as shown at G in Fig. 31. Do the same thing with the other guide strips and support and your cross-head guide is done.

The Cross-Head Guide Block.—This is a block of metal to which the piston rod is secured and which slides in the cross-head guide. To make it take a bar of brass $\frac{5}{8}$ inch square and file out the corners on opposite sides to a depth of $\frac{1}{8}$ inch and on the other sides to a depth of $\frac{1}{16}$ inch as shown at K in Fig. 31.

It must slide to and fro the length of the guide easily and yet without the slightest play. A

A 1/24-H. P. Horizontal Steam Engine

cross section view of the cross-head guide and the guide block is shown in Fig. 35. Drill a ⅛-inch hole through it, thread it with an 8–32 tap and screw it to the free end of the piston rod.

The Rocker Arm.—This is to connect the slide valve stem with the eccentric rod. Cut off a piece of brass 3/32 inch thick, 5/32 inch wide and ⅞

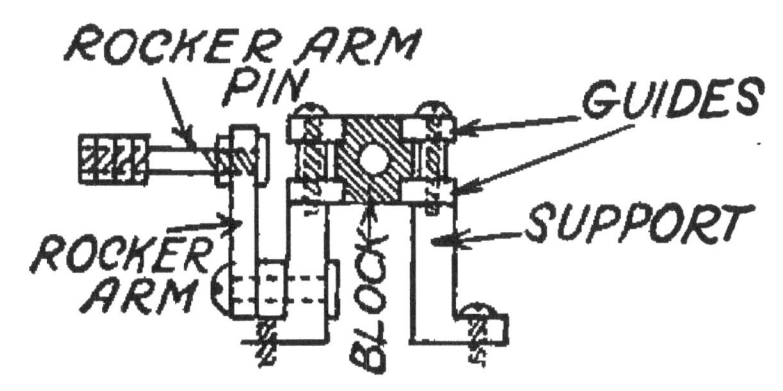

Fig. 35. The Rocker Arm and Cross-head Guide

inch long; file it down until it is ¼ inch wide at one end; file the ends round and drill two ⅛-inch holes, 9/16 inch apart, as shown in Figs. G 31, 34 and 35.

Make a brass washer ⅛ inch thick and ¼ inch in diameter with a ⅛-inch hole in it; slip a ⅝-inch long 8–32 screw through the hole in the large end of the rocker arm, put on the washer, set the screw through the hole in the guide support and screw a couple of nuts on it.

Cut off a piece of steel rod $1/8$ inch in diameter and $7/8$ inch long for the *rocker arm pin* and thread both ends of it; screw a nut on one end, $5/16$ inch down, put this end through the end in the upper end of the rocker arm and screw a nut on it as shown at G in Fig. 31 and in Fig. 35.

The Valve Stem and Bearing.—The valve stem, see Figs. 32 and 34, is a steel rod $3/32$ inch in diameter and $3\frac{1}{8}$ inches long threaded on both ends. Make a bearing to connect the valve stem to the rocker arm by filing down a piece of brass rod, $1/4$ inch in diameter and $9/16$ inch long, on both sides until it is $1/8$ inch thick for $5/32$ of an inch of its length.

Drill a $3/32$-inch hole in the large end and thread it to fit the end of the valve stem; drill a $1/8$-inch hole through the flat end and file it out with a half-round jeweller's file so that when it is on the pin of the rocker arm it can move to and fro in a straight line without binding and at the same time without any lost motion. It is shown at L in Fig. 31.

The Connecting Rod.—This can be made of a

straight strip of brass ⅛ inch thick, ⁵⁄₁₆ inch wide and 4⅝ inches long. Drill a ³⁄₃₂-inch hole in one end and a ⅛-inch hole in the other end. The centers of the holes should be exactly 4⁵⁄₁₆ inches apart. The connecting rod is shown coupled to the cross-head guide block and to the crank in the side view Fig. 36.

The Connecting Rod Bearing.—The connecting rod is coupled to the cross-head guide block by a bearing made of a piece of brass rod ⁷⁄₃₂ inch square and ½ inch long. File down one end until it is ⅛ inch in diameter for ³⁄₁₆ inch of its length and thread it.

Drill a ³⁄₃₂-inch hole through the other end and saw and file a slot ³⁄₃₂ inch wide and ¼ inch deep as shown at M in Fig. 31 so that the thin end of the connecting rod will fit in it. Put a pin through the bearing and the connecting rod and, while it must fit the bearing tight, the connecting rod must swing easily on it.

The Crank and Crankshaft.—Drill a ⁵⁄₃₂-inch hole through the large end of the crank casting and a ⅛-inch hole through the small end and have these holes exactly ⅝ inch apart between their centers as shown at M in Fig. 31.

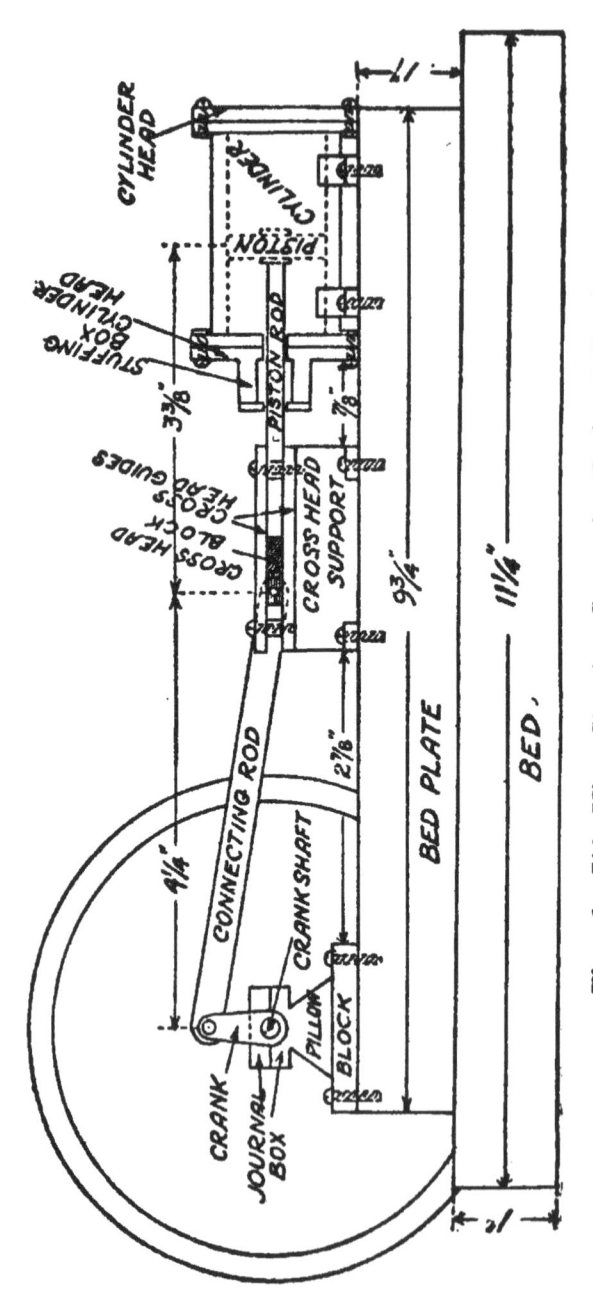

Fig. 36. Side View Showing Connecting Rod and Crank

Drill a 1/16-inch hole through the large end of the crank at right angles to and through it, until it meets the hole drilled for the crankshaft; thread the hole and put in a screw pointed at the end for the set screw. Now put a 1/8-inch thick bolt through the connecting rod and the crank and screw on a nut.

For the crankshaft get a piece of soft steel rod 5/32 inch in diameter and 2½ inches long. Start a drill hole 1/8 inch from one end, another 3/16 inch from the other end and a third hole 1 3/16 inch from the first end of the crankshaft but on the other side of it—that is the first and second holes will be 10 degrees apart as shown at N. Drill these holes about 1/32 of an inch deep to form cavities for the pointed set screws.

Fit the crank on the end of the crankshaft that you drilled first and screw in the screw until the pointed end of the latter sets in the cavity in the shaft and this will hold them securely together.

The Eccentric.—Turn on a lathe, or file out by hand, a groove 1/16 inch deep and 1/4 inch wide in the rim of the eccentric as shown at I; drill a 3/16-inch hole exactly 1/4 inch from the center of the disk and drill a 1/16-inch hole through the groove at the point where it is nearest the shaft hole until they meet.

Countersink the small hole, thread it and put in a flat head machine screw with a pointed end. Slip the eccentric over the shaft so that the pointed screw is directly over the cavity nearest the middle of the shaft and force the screw in until the eccentric is firmly fixed to the shaft and the head of the screw is flush with the face of the groove.

To make the *strap* for the eccentric take a strip of brass $1/16$ inch thick, $1/4$ inch wide and $3\tfrac{1}{8}$ inches long; drill a $3/32$-inch hole in each end, then bend the strip of brass on an iron rod about $1/2$ inch in diameter, hammering it into a ring with a wooden mallet, and bend out the ends as shown at I. Now slip the strap into the grooved eccentric and you are ready to make

The Eccentric Rod.—This is simply a strip of brass $3/32$ inch thick, $1/4$ inch wide and $3\tfrac{3}{4}$ inches long. Drill a $1/8$-inch hole in one end to fit the rocker arm pin and saw a slot $3/32$ inch wide and $1/4$ inch deep in the other end.

Cut out a piece of brass $3/32$ inch thick, $1/4$ inch wide and $3/4$ inch long; drill a $3/32$-inch hole in one end of this piece and saw a slot $3/32$ inch wide and $1/4$ inch deep in the other end. Now slip the slotted ends of the two pieces of metal together as shown at P and solder them.

Place the end between the ends of the eccentric strap, put a screw through the holes of all of them and screw on a nut, all of which is shown in Fig. 34. The strap must fit into the groove of the eccentric to prevent any play but it must not bind. Slip the other end of the eccentric rod over the rocker arm pin and screw on a nut.

The Flywheel.—Should you have the flywheel cast, or buy it in the rough, you must *face* it, or have it faced, in a lathe—that is the rough surface must be cut off with a turning tool.

Then bore a $3/16$-inch hole through the exact center of it so that it is perfectly balanced. Drill a $3/32$-inch hole through the hub of the wheel, thread it and put in a pointed machine screw. Adjust the flywheel on the end of the shaft and screw up the set screw.

An iron flywheel casting can be bought for 25 cents, as I have said before, and a brass one can be had for about 75 cents. It will cost an additional 50 or 75 cents to have a hole bored in the wheel for the shaft and to face it.

The Pillow Blocks.—If these are cast in brass, they will make very good bearings. Drill a ⅛-inch hole in each foot and drill a 3/16-inch hole in each end of the cover and on through into the block to a depth of ⅜ inch; thread the latter holes in the block and ream out the holes in the cover to a diameter of ⅛ inch for the screws.

Now screw the covers to the blocks tight, and drill a 3/16-inch hole through each cover and block to form bearings as shown at J. The centers of these holes must be precisely ⅝ inch from the base of the block and they must be drilled true for, if the holes are the slightest bit out of line the shaft will bind.

To keep the shaft in place when it is set in the pillow blocks get, or make, two brass collars, or washers will do, ⅛ inch thick, ⅜ inch in diameter and with a 3/16-inch hole in each one. These collars must be put on over the ends of the crankshaft before the crank and the flywheel are fixed to it.

Mounting the Engine on the Bed Plate.—The

A 1/24-H. P. Horizontal Steam Engine

top of the iron bed plate should be *planed* off, but if this means an outlay of too much money, then file down those parts of it where the cylinder and steam chest, the cross-head guide and the pillow blocks are to rest on it with a mill file.

Now comes a ticklish job—the drilling of the bed plate to correspond to the holes in the parts above named so that they can be bolted down to it. The safest way to do this is to mark out a sheet of tin the exact size of the bed plate, set the parts on it in the positions shown in the top and side views, Figs. 33, 34 and 36, and *scribe* the positions of the holes on it.

Punch holes through the *template,* as a tin pattern is called, with a center punch, lay it on top of the bed and make corresponding dents in the latter by hitting the center punch with a hammer.

Next drill all holes with a 1/8-inch drill and also drill a 3/16-inch hole through each foot. You can now bolt the parts to the bed plate and feel reasonably sure that each one will be in exactly the right place.

Setting the Engine on Its Bed.—To raise the

engine from the surface it is to set on high enough for the flywheel to clear it the bed plate must be set on a bed.

This can be made of a block of wood 1 inch thick, 4½ inches wide and about 12 inches long. One corner must be cut out to allow a free space for the flywheel to run in. Finally, screw the

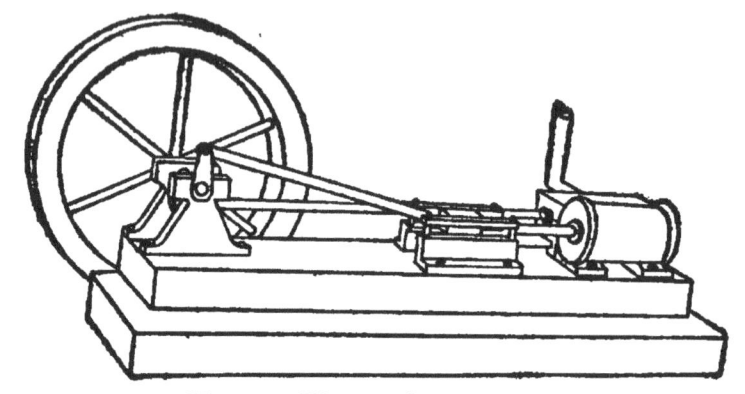

Fig. 37. The Engine Complete

bed plate to the bed and your engine is finished and all ready to run. It is shown as it looks when done in Fig. 37.

The Auxiliary Parts.—The steam pipe and fittings and the *governor* for this engine will be described in Chapter VII, *"Auxiliary Parts for the Steam Engine."* Specifications for the boiler to run it will be found in Chapter V, *"On Making Small Boilers,"* while the safety valve, steam gauge, gauge cocks, whistle, etc., will be described and pictured in Chapter VII.

How the Engine Works.—To understand how a slide valve engine works look at the cross-section view shown in Fig. 33 and imagine that the steam chest is connected to a boiler of the right size to run it.

Now as far as the mechanical operation of the engine is concerned it works like this: The steam from the boiler is forced under pressure into the steam chest and then on through one of the inlet ports into the cylinder.

Since the piston is at this end, the steam pressing against it pushes it to the opposite end of the cylinder; as the piston moves back it does two things, namely: (1) it pushes the used steam up through the other port and into the slide valve which at this instant also opens into the exhaust port and, hence, the steam passes out and into the open air, (2) it pulls the crank half way around and this, of course, turns the crankshaft with it.

The eccentric, which is fixed to the crankshaft, of course, turns half way around too, but it is so placed on the shaft—180 degrees from the crank—that its operation moves the slide valve, to which it is connected by the eccentric rod and slide valve stem, in the opposite direction to that of the moving piston.

The instant the slide valve slides across the slide valve plate it connects the first inlet port with the exhaust port; at the same time it opens the other intake port from the steam chest into the cylinder and the piston is forced the other way. In this fashion the steam acts first on one side of the piston and then on the other, driving it forth and back.

The purpose of the flywheel is to carry the crank around past the *dead centers;* that is the points where the piston reverses its direction and consequently where the steam ceases to act on it. For the same reason the flywheel, due to its *inertia,* as it is called, steadies the motion of the engine. How the steam in the cylinder acts on the piston and forces it from one end of the cylinder to the other will be described in Chapter VIII, on *"How Steam Works."*

CHAPTER V

MAKING SMALL BOILERS

A $1/12$ Horse Power Vertical Tube Boiler: A Simple Iron Boiler: The Shell, The Smoke Box, The Firebox: The Grate; Gas and Liquid Fuel Burners; A Good Copper Boiler: The Shell, The Smokebox, The Firebox, Fittings for the Boiler—How to Test the Boiler—A Safe Way to Operate a Small Boiler.

In this chapter I shall tell you how to make two small boilers. The first is a *vertical boiler* large enough to run the steam turbine, described in Chapter II, or the horizontal engine, described in the last chapter, at full speed.

A boiler should be at least twice the horsepower of the engine it is to run. Now you can make a boiler according to the plans and specifications I have given or you can figure one out according to the rules laid down at the end of this chapter and design and construct it to suit yourself.

A $1/12$-Horse Power Vertical Tube Boiler.— This boiler is of the single tube type, that is it has a *fire-tube,* or flue, running up through the middle of it so that the fire not only acts on the *fire-box sheet,* as the end of the boiler next to the fire is called, but it also heats the tube, and hence the water around it, as the burning gases pass through it to the smoke stack.

It takes a steam pressure of about 8 or 10 pounds to run the $\frac{1}{24}$ h.p. engine, and a boiler 6 inches in diameter and not less than 8 inches high must be used, for, if it is any smaller, it will not make steam fast enough. Better make it a third or a half larger than this size and be sure of a continuous performance.

There are a couple of simple ways to make a small boiler and I will explain and picture both of them and you can take your choice.

A Simple Iron Boiler—*The Shell.*—Have a steamfitter cut a piece of iron pipe for you $6\frac{1}{4}$ inches in diameter, outside measurement, and 9 inches long, and thread both ends of it.

Also have him fit a cap to each end of the pipe and bore a 1-inch hole through the center of each one. Now drill, or have drilled, a $\frac{1}{4}$-inch hole on each side of and 2 inches from the center hole in the top cap as shown in the cross section view

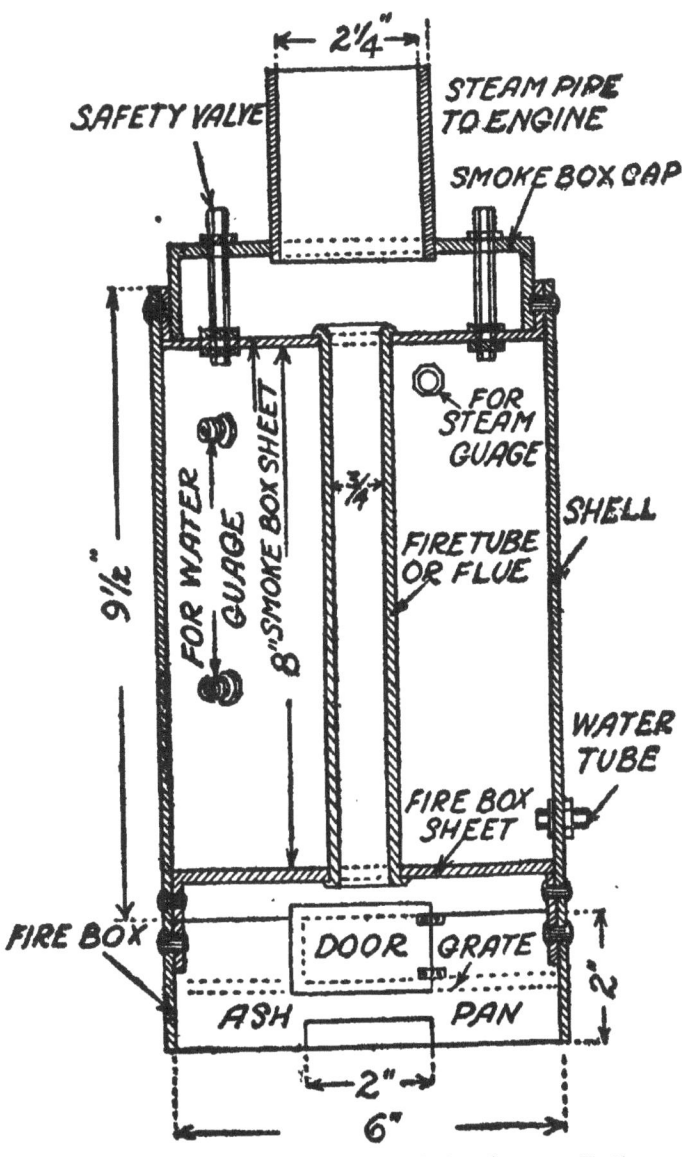

Fig. 38. Cross-section View of the Copper Boiler

94 The Boys' Book of Engine-Building

Fig. 38, and the top view, Fig. 39, and thread them.

Cut off two pieces of ⁵⁄₁₆ inch pipe 2½ or 3 inches long and thread both ends of them. Screw a nut on each end of each pipe, smear the

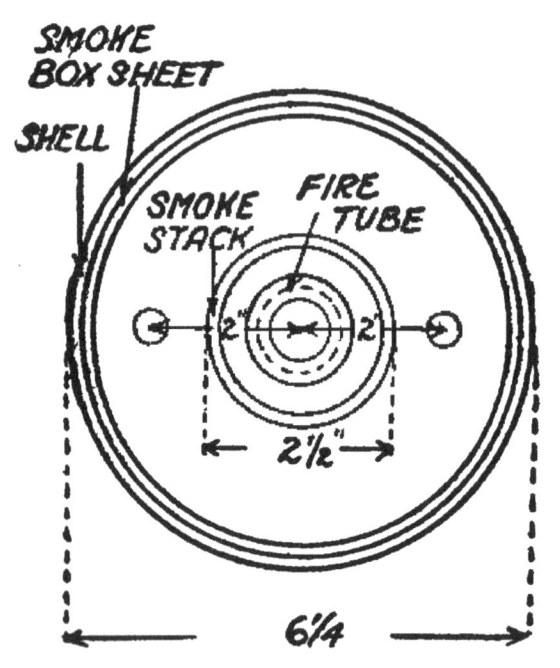

Fig. 39. Top View of the Boiler

ends with red lead, screw them into the holes in the cap and then screw another nut on the end of each pipe, as shown in Fig. 38. This done, screw both caps on the ends of the large pipe good and tight.

Next get a *seamless copper tube* [1] 1 inch in diameter and 9¼ inches long; put it through the holes in the caps, or *fire-box* and *smoke-box sheets* as they are now called, so that it projects through each one ⅛ inch. This done the ends of the *flue,* as a single fire-tube is usually called, must be *expanded,* that is spread out all around to make it fit steamtight as shown in Fig. 38.

To do this get a piece of iron rod 1 inch in diameter and 3 inches long and turn it down in a lathe, or file it down until it tapers enough so that the small end will easily go into the tube; now drive this tool, or *swage,* as it is called, see A, Fig. 40, into and pull it out of an end of the tube until the latter is made larger, or expanded, when the tube will crowd up close to the sheet, or cap, and form a steamtight joint. It is a good plan to run some solder around the joint to further insure its being tight.

Make a peening tool of a bar of iron say ¼ inch square and 3 inches long—the exact size doesn't matter—and file a notch in one end as

[1] Can be bought of V. T. Hungerford Brass and Copper Co., 80 Lafayette St., New York, or of Patterson Brothers, 27 Park Row, New York City.

shown at B. Set this tool with the notch on the rim of the tube and pointed outward; then with the aid of your hammer gently turn the rim down all around until it laps over the sheet and fits tight up to it. Do the same thing with the other end of the tube and the boiler *shell* is done.

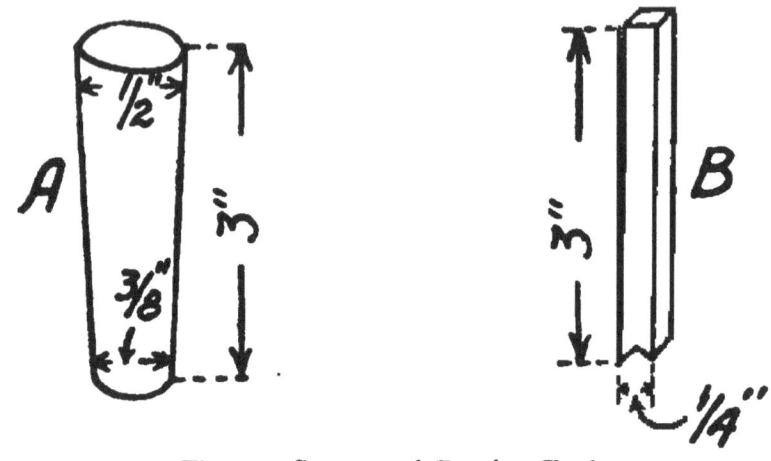

Fig. 40. Swage and Peening Tools

Drill ¼-inch holes in the boiler shell at the places shown in the cross-section view, Fig. 38, and thread them to fit ¼-inch iron piping. Then screw a piece of ¼-inch pipe, 1 inch long, threaded at both ends and smeared with red lead into each hole. The intake water pipe, water gauge and steam gauge pipe are to be coupled to these pipes.

The Smoke-Box.—Get a cap 5½ inches in diameter, bore a hole 2¼ inches in diameter through its center and have it threaded; also get a pipe 2½ inches in diameter, 3 or 4 inches long, for the smokestack and threaded on one end to fit the hole in the cap.

Drill two ¼-inch holes in the smoke-box cap, 2 inches from the center of the large hole, when the cap will slip over the pipes screwed in the smoke-box sheet as shown in Fig. 38. Screw a nut on the end of each pipe to hold the cap in place and then screw the end of the smoke stack into the cap.

You can make a conical smoke-box, a flanged smokestack and a flaring fire-box by turning patterns of these parts in wood and having them cast in brass or iron. A boiler made in this way will look better but it will not work any better.

The Fire-Box.—To make the fire-box take an iron, brass or copper ring just large enough to fit around the lower cap—I should say fire-box sheet—and 4½ inches high; drill three or more ⅛-inch holes at equal distances apart around one edge and a like number of ³⁄₃₂-inch holes at the same distances apart, and thread these latter holes with an 8-32-tap.

Saw, drill or cut out a piece of metal from the ring on the end having the screw holes 1 inch wide and 2 inches long; this makes an opening for the fire-box door. To make the door take a piece of metal ⅛ inch thick, 1½ inches wide and 2½ inches long; get a pair of small brass hinges and rivet these to the door and to the fire-box rings.

Also saw out with a hack saw, or cut out with a cold chisel, an opening on the bottom of the fire-box ½ an inch wide and 2 inches long for the air draft.

If solid fuels, such as charcoal, coke and the like are to be burned in the fire-box, it must have a *grate,* but if gas, kerosene or alcohol is to be used a grate is not needed though each kind requires a special burner which will be described presently.

The Grate.—To make a grate take a ring of iron ⅛ inch thick, ¾ inch high and cut seven parallel slots ⅛ inch wide and $\frac{3}{16}$ inch deep in one end and drill four holes through it at equal distance so that the grate can be bolted to the shell of the fire-box as shown in Fig. 41; now cut seven strips of either square iron bar or round iron rod ⅛ inch thick and of varying lengths to fit the ring and drop these into the slots. The grate thus formed will fit inside the fire-box and, once

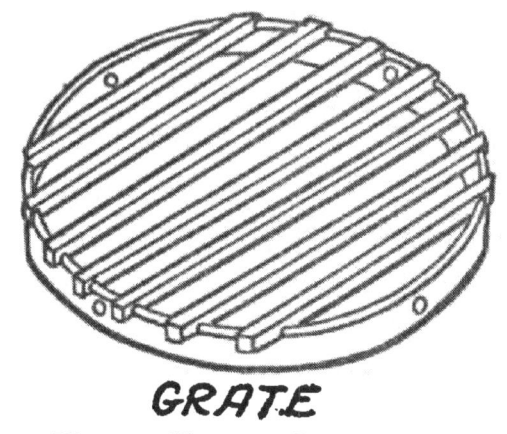

Fig. 41. How the Grate is Made

in, the grate bars cannot slip out. An easier way to make a grate is to make a wood pattern first and have it cast in iron.

Gas and Liquid Fuel Burners—*An Alcohol Burner.*—A burner of this kind is shown at A in Fig. 42. It will serve the purpose well and is the next best thing to a Bunsen burner. It can be bought for 25 cents.

A Gas Burner.—A Bunsen flame burner like the one shown at B in Fig. 42, gives a broad, hot flame and is quite expensive, costing as it does in the neighborhood of $3.00.[1]

A Kerosene Burner.—You can make a burner

Fig. 42 A. An Alcohol Burner

of this kind with little trouble and at a small cost. Cut out of sheet iron a plate $\frac{1}{16}$ inch thick and 7 inches in diameter and turn up the

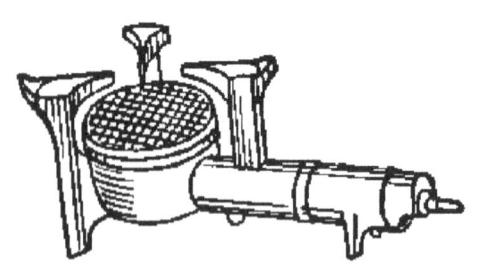

Fig. 42 B. A Bunsen Flame Stove

edges $\frac{1}{8}$ inch all the way round to form a pan; rivet three legs to it so that it can be set under the boiler with the plate at about the height of

[1] L. E. Knott Apparatus Co., Boston, Mass.

Making Small Boilers

the grate and tilting toward the back a little as shown at C in Fig. 42.

The next thing to do is to make a hole in the bottom of a pint kerosene oil can and solder a cock

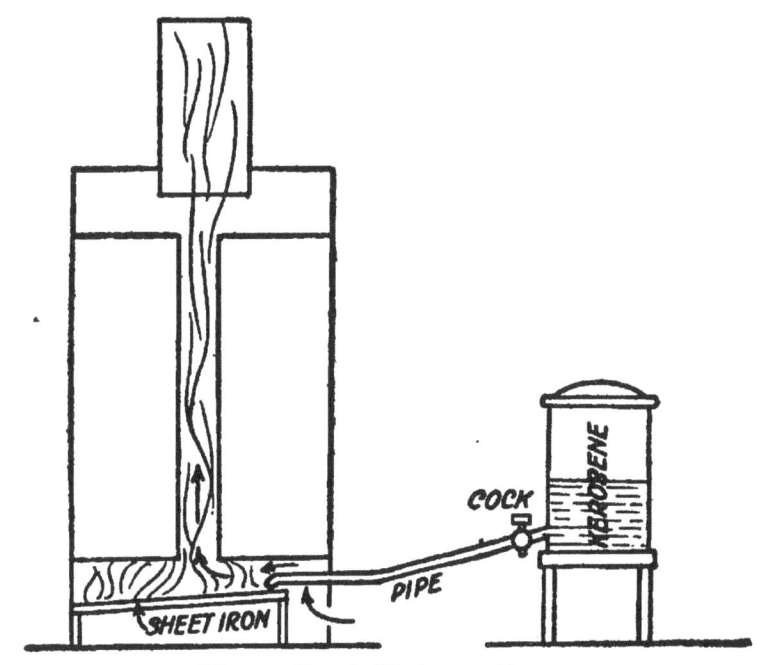

Fig. 42 C. A Kerosene Burner

to it and in turn solder to this a ¼-inch pipe, 12 inches long; cut a ¼-inch hole through the top of the door of the fire-box, pass the pipe through the hole and over the pan, all of which is shown at C.

Turn on the oil so that it falls drop by drop on the pan and let it spread over the whole surface, then light it and it will burn up with a large flame.

A Good Copper Boiler—*The Shell.*—To make this little upright boiler get a piece of *seamless* copper tube with a ⅛-inch thick wall, 6 inches in diameter and 11½ inches long. Brazed copper tube is not nearly so good because it might expand and pull apart; if you have to use it, wrap a layer of No. 16 Brown and Sharpe gauge steel wire around it.

Or, you can get a sheet of soft rolled copper 11½ inches wide and 20 inches long, form it into a cylinder 6 inches in diameter, lap the edges over three-fourths of an inch and rivet it. You must use large rivets tapering from ¼ inch at the end to ⁵⁄₁₆ inch at the head. Drill the holes in the seam 1¹⁄₁₆ inch apart, measured from their centers; put in the rivets and rivet the seam tight.

Whether you use a drawn copper tube or form one yourself, for the boiler *sheets,* as the ends of the boiler are called, use ⅛-inch thick soft rolled copper. Cut out two disks from this sheet and have one of them 7½ inches in diameter for the smoke-box sheet and the other one 8½ inches in diameter for the *fire-box sheet.*

Making Small Boilers

Cut out a 1-inch hole in the center of both sheets. Next, hammer the edge of the small sheet over ¾ inch all round, thus making a pan of it. Likewise bend the edge of the large sheet over in the same way 1½ inches. When the edges of these sheets are hammered into shape, each pan must fit snugly into the ends of the tube.

Put the smoke-box sheet in the boiler tube first, with the bent edge up as shown in Fig. 38 and drill a 5/16-hole through the tube and the rim; slip a rivet through the holes from the inside, hold a large hammer against its head and with a machinist's ball peening hammer, peen the end of the rivet, that is hammer it down with the ball end until it spreads out evenly all round and draws the two pieces of metal together as tightly as possible.

Drill another hole through the rim of the sheet and the tube on the opposite side and rivet them together to hold the sheet in place. Keep on drilling holes with their centers 11/16 inch apart all the way round, then slip a rivet into each hole and hammer it down.

Now set the fire-box sheet in the tube with the

edge of the rim out and projecting ¾ inch beyond the end of the tube, the purpose of which is to allow the fire-box ring to be riveted to it as shown in Fig. 32. These riveted lap seams should not leak steam at a pressure of fifteen or twenty pounds but as a precautionary measure you can run solder into them.

The next thing to do is to put in the *central flue*. This is done in precisely the same way as it is in making the iron boiler previously described. The tube should be 1 inch in diameter and 8½ inches long. Be sure that the ends of the tube are well expanded and don't forget to run solder around the joints before you turn over the edges of the tube.

The Smoke-Box.—If you can get an iron pipe cap 5½ inches in diameter, making the smoke-box becomes a very simple matter. Have a steam-fitter bore a 2½-inch hole in the center of it, thread it and screw in a piece of pipe 4 inches long. The way this is fastened to the smoke-box sheet will be explained further on.

Should you want to make the smoke-box, cut out a disk of copper 8 inches in diameter and hammer the edge of it over all round making it

1¼ inches wide. Cut a hole 2 inches in diameter as before, get a piece of brass or copper tubing 2 inches in diameter and 4 inches long, thread one end of it, screw on a thin brass nut, slip the end through the hole in the smoke-box and screw on another nut on the other side.

A conical cast smoke-box with an overhanging flange will make the boiler look like the real thing, and, if you make a pattern of it and have it cast, make a pattern of the fire-box at the same time and have it cast in the same metal.

The Boiler Connections.—All of the fittings of the boiler, namely the water intake, the steam outlet, the pressure gauge, the water gauge and the safety valve, must be provided for while the boiler is under construction, that is before the sheets are riveted in the ends of the boiler tube.

The way to do this is to drill ¼-inch holes in the tube and smoke-box sheet at the places shown in Figs. 38 and 39; put a ¼-inch pipe, 1 inch long and threaded at both ends into each hole; put a leather washer on each pipe on both sides of the boiler and screw on a nut in and outside.

The Fire-Box.—This is the lower part of a

vertical boiler on which the latter rests and in which the fire is kept to heat the water.

There are several ways to make a fire-box, but the easiest is to get a piece of brass or copper tube the exact diameter of the tube of your boiler and 3 inches high. Saw or cut out an opening for the door and another for the air draft in the manner described for the iron boiler and also a door for the former. Now rivet the ring to the projecting end of the fire-box sheet as shown in Fig. 38.

If you intend to heat the boiler with a solid fuel, you will, of course, need a grate, but if you want to use gas or liquid fuels then select one of the burners I have previously described for it.

Fittings for the Boiler.—Screw a stop-cock on to the water intake pipe; a hand or a power *force pump,* or an *injector* must be coupled to the stop-cock and to a can of water or other source of water supply.

Screw an elbow on to the outlet steam pipe, screw a 1-inch length of pipe into the elbow and screw a stop-cock to the end of the pipe; screw a 2-inch length of pipe into the other end of the stop-cock, screw on an elbow and then screw another length of pipe into the elbow long enough to reach to the engine.

You will need an elbow screwed to the pipe for the pressure gauge and, finally, if you have a steam whistle, screw a coupling on the pipe in the smoke-box sheet. Pipe ¼ inch in diameter can be bought cut to any length up to 2 feet and threaded on both ends, and the elbows and couplings are tapped with right-left-handed threads to fit the pipe. The boiler complete with a cast smoke-box and fire-box is shown in Fig. 43.

How to Test the Boiler.—Before you get up steam in the boiler after you have completed it, the safest way is to test it by trying it with water pressure. Then you can find out where it leaks and why without danger either to yourself or to your handiwork.

If you will turn to the next chapter, you will find that the steam gauge is constructed with a bent tube so that instead of the steam acting directly on the mechanism of the steam gauge the tube is filled with water and the steam presses on the water instead.

Hence all you have to do is to connect the intake water pipe to a force pump and to pump

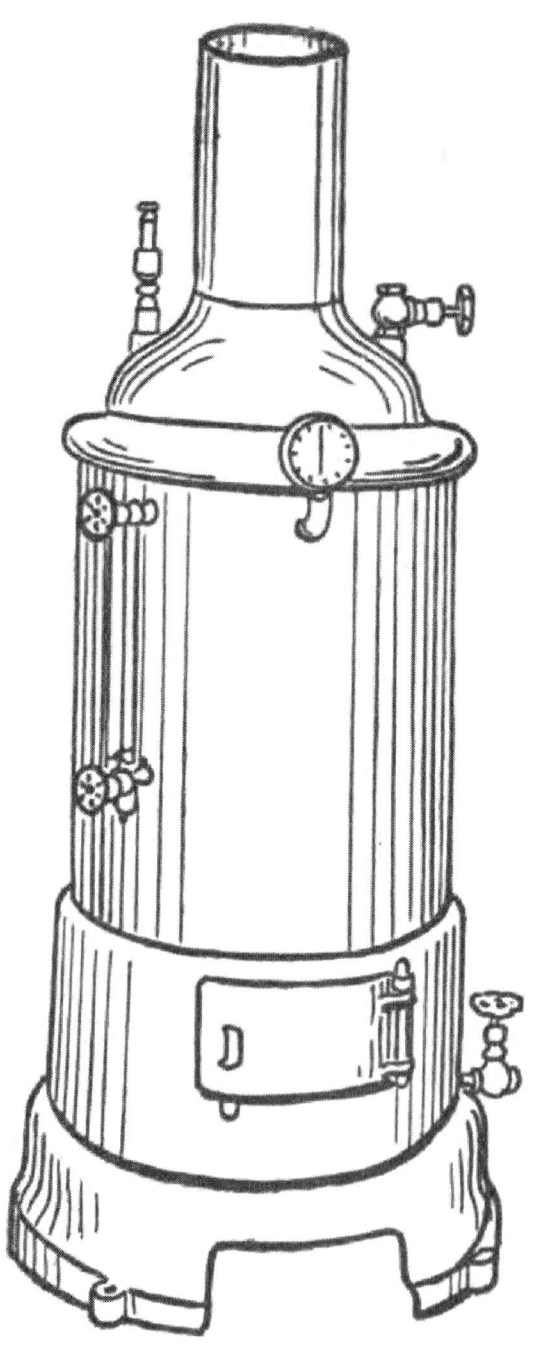

Fig. 43. The Boiler Complete

until the pressure gauge indicates twice the number of pounds pressure that you ever intend to get up in steam. The boilers described in this chapter should be tested to 40 or 50 pounds pressure; consequently you should never let the steam pressure rise to more than 20 pounds.

A Safe Way to Operate a Small Boiler.—Any engine that will run with steam can be run with compressed air.

Now by pumping air into the boiler until it is under as much, or a little more pressure than steam, if you used it, you can run your engine without steam and, naturally, without fire, and this eliminates both of these elements of danger.

To pump up the boiler is a simple matter, for you can do it with a bicycle pump, or, better, with a motor car pump. If you use compressed air, you do not need the water gauge on the boiler; instead of coupling a source of water to the intake pipe fit a bicycle valve or an inner tube valve—which is often thrown away or one can be bought at an auto supply house—in the intake **pipe and then connect the latter with the pump.**

CHAPTER VI

FITTINGS FOR MODEL ENGINES

Pipe and Fittings—Taps and Stopcocks—The Steam Whistle—Safety Valves: The Lever Safety Valve; A Spring, or Pop, Safety Valve—The Governor—A Steam Force Pump—The Injector: How to Make It; How It Works—The Water Gauge—The Steam Gauge: How to Make It; Calibrating the Dial.

In building model engines, especially steam engines, a large number of fittings such as pipe, *unions,* or pipe couplings, *stop-cocks, whistles, safety valves, governors, force pumps, water* and *steam gauges* are needed.

While it is usually more convenient to buy them ready-made of the right size to put on your engine, or boiler, it may happen that you would rather make them yourself and so I'll tell you how.

Pipe and Fittings.—You can buy brass tubing of any size you want from $\frac{1}{16}$ inch in diameter on up and cut to any length you may need, but it is not always an easy matter to get couplings,

Fittings for Model Engines

elbows, T's and crosses that are threaded to fit.

The Weeden Mfg. Co., of New Bedford, Mass., sells pipe and fittings in small sizes, threaded and ready to use; you can buy brass or copper tubing in nearly all sizes from Patterson Brothers, 27 Park Row, New York City, and

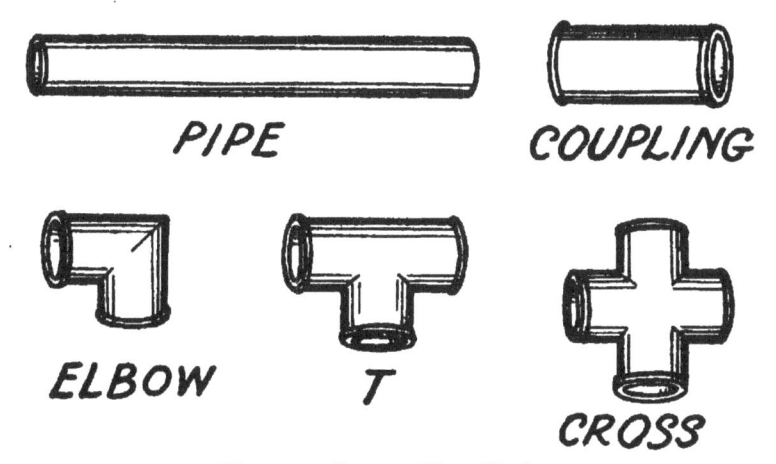

Fig. 44. Steam Pipe Fittings

elbows, T's and crosses made of brass ⅛ inch and ³⁄₁₆ inch in diameter are sold under the name of *brass sockets* by the Ideal Aeroplane and Supply Co., 82–86 West Broadway, New York City, and the Wading River Mfg. Co., Wading River, Long Island, N. Y.

Regular steam pipe and fittings like those shown in Fig. 44 can be bought in ¼ inch and

$\frac{5}{16}$ inch sizes, outside diameter, of Luther H. Wightman and Co., 132 Milk Street, Boston, Mass., and the Chicago Model Works, 166 West Madison Street, Chicago, Ill.

Taps and Stop-cocks.—A small brass tap or stop-cock half size shown at A in Fig. 45 can be bought for 50 cents, a larger one for 65 cents and a still larger one for 75 cents.

A good way to make a stop-cock is to cut off

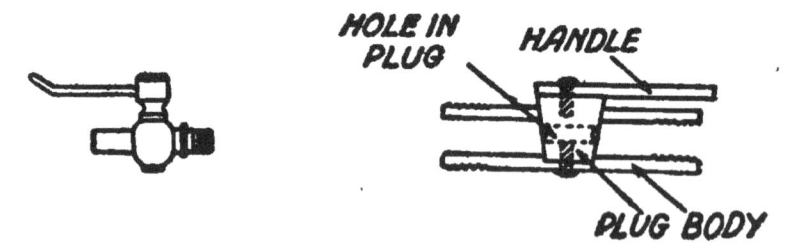

Fig. 45. A Brass Stop-cock. (Full Size.)

a piece of brass rod the size you want it and drill a hole through it from end to end, the diameter of which should be about half the diameter of the rod.

Next drill a *plug hole* for the *plug* through the middle of the stop-cock body as shown in the cross section drawing at B in Fig. 45, *ream* it out to make it slightly conical, and drill a hole through the body opposite the plug hole for a screw. The plug, which is a conical piece of brass, must fit into the plug hole nicely and have a hole drilled through it as is also shown at B. Drill a 1/16-inch hole in each end and thread them.

The next operation is to grind the plug to make it fit the plug hole steamtight; to do this make a grinding paste of fine emery mixed with a few drops of sewing machine oil and thinned down with kerosene. Rub the mixture on the conical edge of the plug, then set the plug in the plug hole and turn it around forth and back but don't press too hard on it.

When the plug is *seated,* as it is called, push a screw through the body of the stop-cock and screw it into the bottom of the plug. Now make a handle for it and screw it to the top of the plug and your stop-cock is done.

The Steam Whistle.—The kind of a steam whistle used on a locomotive is called a *bell whistle* because it is fitted with a hollow top closed at the upper end and which somewhat resembles a bell.

The first thing you need to make a steam whistle with is a stop-cock like the one just de-

114 *The Boys' Book of Engine-Building*

scribed. For a whistle of the right size for a model locomotive make a stop-cock of a piece of rod having an outside diameter of ⅛ inch and drill a ¹⁄₁₆-inch hole through it.

Drill a ¹⁄₁₆-inch hole clear through one end of the plug body and another at right angles to it,

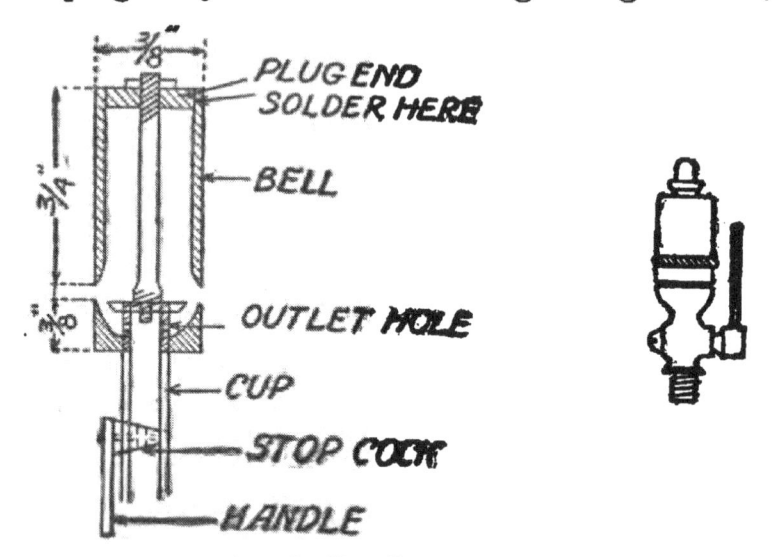

Fig. 46. The Steam Whistle

so that the end will have four outlet holes for the steam to escape through. Put the plug in the middle of the body as before and thread both ends of the latter.

Make a cup of a disk of brass ³⁄₁₆ inch thick and ¾ inch in diameter and hollow out one side as shown in the cross-section at A in Fig. 46,

Fittings for Model Engines 115

either with the point of a large drill or with a countersink. Drill a $3/32$-inch hole through it and thread it to fit the end of the stop-cock body; screw the cup on the end with the *concave,* that is the hollow part up, and screw it down far enough so that the holes in the end of the tube are just above the hollow of the cup.

Cut out a disk of sheet brass $1/32$ inch thick and $1/4$ inch in diameter, drill a $1/8$-inch hole through the center of it, thread it and screw it to the end of the stop-cock body so that it sets flush with the edge of the cup. Now thread the inside of the end of the body and screw in a brass rod $3/32$ inch thick and $7/8$ inch long but not in far enough to stop up the outlet holes in the end of the stop-cock.

For the bell of the whistle cut off a piece of brass tube $3/8$ inch in diameter and $3/4$ inch long; make a plug end for it of a piece of brass rod and either screw, or solder, it into one end; drill a $1/16$-inch hole through it and thread it; then thread the end of the brass rod and screw the latter into it and, finally, put a nut on the end to hold the bell on tight. The lower edge of the bell should come within $1/16$ inch of the edge of the cup. By screwing the bell up or down you can adjust it so that it will whistle the loudest.

Now when the handle of the stop-cock is turned on, the steam rushes out of the holes in the end of the body; this fills the hollow cup under the disk when it is forced out and up against the edge of the bell; on striking it the bell is set into violent vibration and a characteristic steam whistle results.

Safety Valves.—There are several kinds of safety valves but I shall describe only two of them and these are: (1) the *lever safety valve* and (2) the *spring safety valve*.

The Lever Safety Valve.—The time-honored lever safety valve is largely used on stationary boilers. The easiest way to make a safety valve of this kind is to take a ¼-inch pipe, 1 inch long, and thread both ends of it.

Cut or saw out two brass disks each $3/16$ inch thick and ⅞ inch in diameter; drill a $5/16$-inch hole through one of them and thread it to fit the pipe; drill a $3/16$-inch hole through the center of the other disk and make it $5/16$ inch in diameter on the other side.

Put the two disks together and drill a $3/32$-inch

Fittings for Model Engines

hole on each side of the holes in the center and thread them; screw the disks together on one side and then make a standard of a piece of brass rod ⅛ inch in diameter and ¾ inch long; thread one end, drill a 1/16-inch hole through the other end and saw a slot 1/16 inch wide and 3/16 inch deep in it; these things done, screw the standard

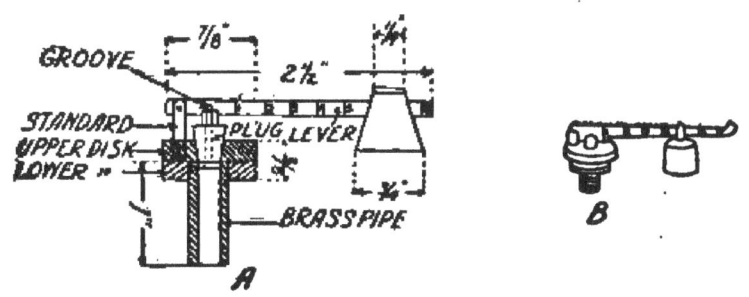

Fig. 47. A Lever Safety Valve

through the disks as shown at B in Fig. 47 and screw the disks together on the other side.

Make a lever 1/16 inch thick, ⅛ inch wide and 2½ inches long; drill a hole in one end of it and set the end of the lever in the slotted end of the standard and pivot it with a pin.

The plug can be made of lead, type metal or brass; it must be a shade larger than ¼ inch at the bottom and a trifle larger than 5/16 inch at the top. If it is made of lead, or type metal, you can fit it into the valve hole steamtight by pressure and turning it around, but if it is of brass you will have to grind it with fine emery mixed with oil.

Drill a $\frac{1}{16}$-inch hole through the center and push a brass wire $\frac{5}{8}$ inches long and threaded on both ends through it; screw a nut on both ends and file a groove in the top end of the wire so that the lever will rest in it.

All that is left to be done is to make the weight; this can be a piece of rod $\frac{1}{2}$ inch in diameter and $\frac{5}{8}$ inch long; saw a slot $\frac{1}{16}$ inch wide and $\frac{1}{4}$ inch deep and put a pin through the upper end. Slip the weight on the lever and screw the safety valve into the boiler shell.

A Spring or Pop Safety Valve.—A spring or pop safety valve is used on locomotives and you can easily make one for your model. If you will take a look at Fig. 53, which shows a cross-section view of the locomotive boiler, you will see that the safety valve pipe which leads through the steam dome outside is $\frac{1}{8}$ inch in diameter and the end of this pipe must be threaded.

Cut out a disk $\frac{1}{4}$ inch in diameter and $\frac{1}{8}$ inch thick; drill a $\frac{3}{32}$-inch hole through its center and

Fittings for Model Engines

thread it to fit the pipe. Cut out another disk of the same thickness and size, drill a ⅛-inch hole in its center and ream it out until it is 5⁄16 inch in diameter on the other side. Make and grind a brass plug to fit this conical hole as shown in Fig. 48.

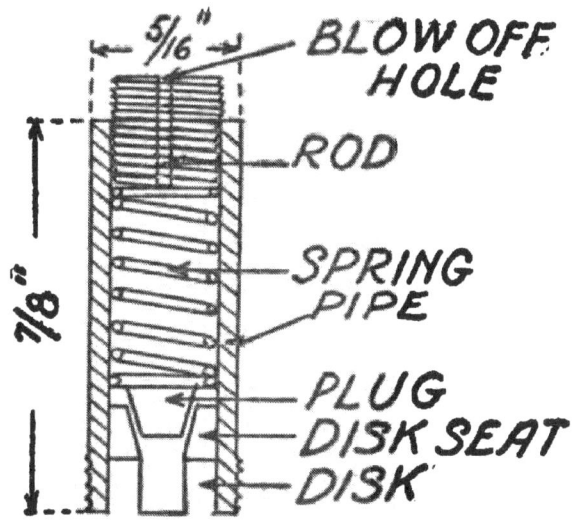

Fig. 48. A Spring or Pop Safety Valve

Fit both of these disks into one end of a pipe 5⁄16 inch in diameter and ⅝ inch long with the valve disk inside, and solder both of them in the pipe. Put the plug in the valve hole; form a spiral spring of brass wire and put it in the pipe on top of the plug; thread the inside end of the pipe and thread a rod ¼ inch long to fit it; drill a ⅛-inch hole through the rod from end to end to let the steam escape when the valve is opened, and cut a slot across one end of this plug and then screw it into the end of the pipe.

The action of this safety valve is simple enough: When the pressure of the steam is higher than it should be it forces the plug up against the pressure of the spring and the steam blows off through the hole in the screw plug at the top of the safety valve.

The Governor.— A *governor* is used on stationary engines where it is necessary for the speed to be *constant*, that is perfectly steady.

The usual kind of governor works on the principle of *centrifugal force,* that is two balls, each of which is fitted to the end of a lever, are made to revolve by the engine; as the speed of the engine increases the balls fly apart and raise the levers. The levers are in turn connected to the throttle valve and when the speed is great enough the raising levers close the valve, the steam is shut off and the speed of the engine is reduced.

To make a *centrifugal governor,* or *ball governor,* or *fly ball governor,* as it is variously called, get a soft steel rod ⅛ inch in diameter

Fittings for Model Engines

and 2½ inches long for the *spindle;* file one end of it flat and drill a 1/16-inch hole through it as shown at A, Fig. 49.

Cut two strips of brass 1/16 inch thick, 3/16 inch wide and ½ an inch long for the *top bars;* lay the bars together and drill a 1/16-inch hole through them at both ends and in the middle; put a bar

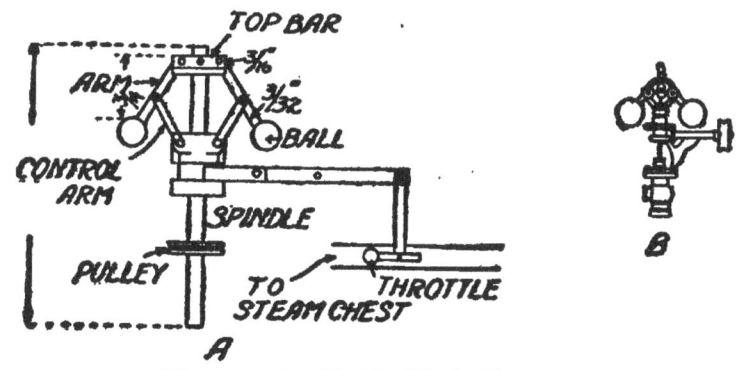

Fig. 49. An Easily Made Governor

on each side of the flat end of the spindle and drive a pin through all of them, tight.

Make two *fly arms* of 3/32-inch brass or steel rod, ¾ inch long; thread one end of each one and file both of them a little flat in the middle and at the plain ends and then drill a 1/16-inch hole through the middle and ends.

Next cut out four flat *control arms* 1/16 inch thick, 3/16 inch wide and 5/8 inch long; lay all of the arms together and drill a hole through the ends of them. Set the ends of a pair of them on the opposite sides of the fly arms and pivot them with pins so that the joints will move freely.

The next thing is the *guide*. Turn or file down a rod ⅜ inch in diameter and ½ inch long, and drill a hole ⅛ inch in diameter through it from end to end so that it will slide easily on the spindle; cut or file a groove, ⅛ inch wide ⅛ inch from the one end; file down the opposite sides of the other end and drill a hole through each side.

Slip the guide on the spindle and joint the flat sides of it to the lower end of the control arms on each side. Screw a ball ¼ inch in diameter to the free end of each of the fly arms and then fix a grooved pulley ½ inch in diameter on the spindle.

Make a *lever* of a strip of 1/16-inch brass 3/16 inch wide and 1¼ inches long, and drill a hole in each end and a third hole ½ an inch from one end as shown at A. Cut out two strips of brass each ¾ inch long of the same thickness and width as the lever; drill a hole in one end of the lever between them and rivet the three pieces together;

Fittings for Model Engines

now bend the ends out until they fit into the groove of the guide.

The free end of the lever is pivoted to another and shorter lever and this in turn is pivoted to another and a still shorter lever and this latter lever is pivoted in turn to the handle of the stop-cock, or *throttle,* as it is called when it is used to open and close a steam pipe, all of which is shown in Fig. 49.

Finally, the lower end of the spindle must rest in a bearing and turn easily, and the grooved wheel is belted to the pulley on the crankshaft of the engine.

Now when you turn on the steam the balls are down and the throttle is open; as the engine gets up speed the balls fly apart, this pulls the guide up and so raises the yoked lever; this shuts off the steam until the speed falls off when the balls begin to drop and more steam is admitted to the engine. A bought governor is shown at B.

A Steam Force Pump.—For stationary boilers a *steam force pump* is generally used to pump water into it as needed.

A simple force pump can be made of a couple of *check valves,* as shown at A in Fig. 50, that an upright position. A small force pump as shown at C can bought for $3.50.[1]

It works like this: When the piston is raised it pulls the air out of the check valves; this pulls the ball in the right hand valve down tighter, but it lifts the ball in the left hand check valve and hence the air and water in the pipe under it is raised up into the barrel in which the piston works.

When the piston is forced down, the pressure of the water presses the ball in the left-hand check valve down tight, but it pushes the ball in the right-hand valve up and this allows the water to pass through it and into the boiler.

The Injector—*How to Make It.*—To feed water into a steam boiler without having to pump it an *injector* is used. While nearly all stationary boilers are fitted with pumps, locomotive boilers are provided with injectors which have no moving parts.

An injector is so made that steam from the boiler is forced through a nozzle, thus forming a jet, and blows the water into the boiler against

---

[1] Can be bought of Luther H. Wightman, Boston, or of the Chicago Model Works, Chicago.

its own pressure. Hence we have what is called a *hydrostatic paradox*.

Get a piece of pipe ⅝ inch in diameter and 2 inches long, thread both ends of it and drill a ¼-inch hole in the pipe ½ inch from one end for the overflow pipe. Fit a cap to each end of the pipe, drill a ¼-inch hole through the center of each one and thread them to fit a ⅜-inch pipe; these are for the steam nozzle and the valve chamber as shown in Fig. 51.

Next you will need an elbow that will fit a ⅜-inch pipe and a piece of pipe ⅜ inch in diameter and ½ or 1 inch long and threaded at both ends for the overflow pipe.

Now make the nozzles and this is no easy job either way you do it. The first way is to take a brass rod ⅜ inch in diameter and cut off three pieces; have one of them 1 inch long, the second 1¼ inches long and the last 1⅞ inches long.

Drill a ¼-inch hole, ½ an inch down in the end of each piece, and a ¹⁄₁₆-inch hole the rest of the ways through; then ream out the holes to form conical chambers in them. File, or, better, turn down in a lathe, the ends of the nozzles to give them a conical form.

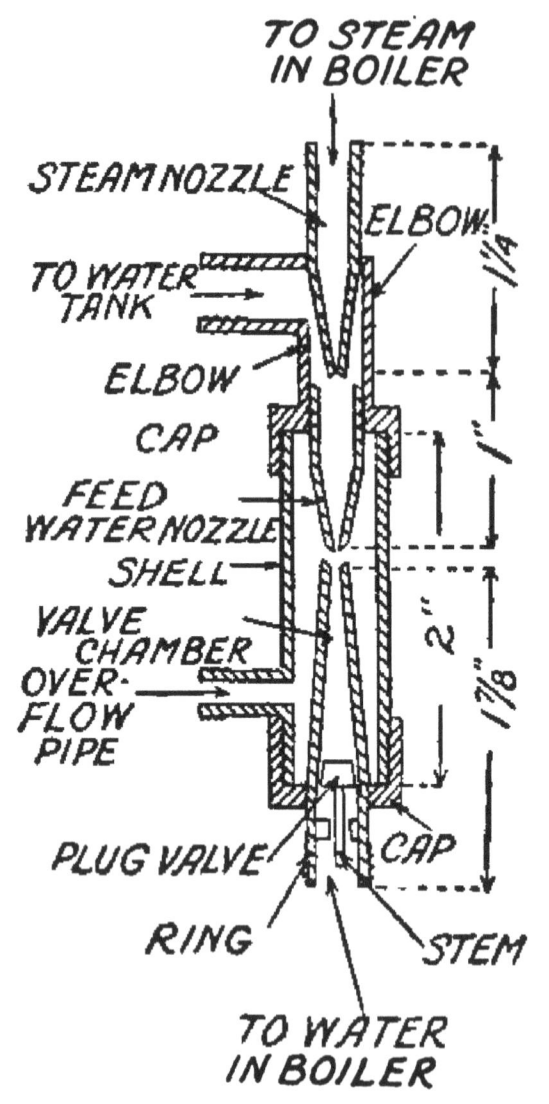

Fig. 51. Cross-section of a Steam Injector

Fittings for Model Engines 129

Another and easier way is to shape up the nozzles of 1/16-inch sheet copper and *hard* solder [1] the seams. Whichever way you make them thread the large ends of the nozzles. This done, make a conical plug that will fit into the valve chamber, screw a stem into it and set it into the lower end; put in a ring with the hole over the stem and solder it there.

Having finished all the parts of the injector the next thing is to assemble it. Screw the feed water nozzle into the cap; screw the elbow on the end of the nozzle and then screw the steam nozzle into the elbow so that the small hole in it is just inside the large hole in the feed water nozzle. Screw the valve chamber nozzle into the end of the other cap; screw both caps on the large pipes and, finally, screw the overflow pipe into the shell.

Now, couple the steam nozzle to the boiler above the water line, connect the valve chamber to the boiler below the water line and join the elbow to the water tank. When all these things are done, if you have not less than forty pounds of steam in the boiler, the pressure of the steam

[1] Borax mixed with water to the consistency of paste.

in the boiler will force the water into the boiler against the back pressure and this is called a hydrostatic paradox.

How it Works.—When the steam flows into the steam nozzle it blows out of the small end in a jet and on down through the feed water nozzle—that is, it does so at first.

The jet of steam forms a vacuum in the elbow and this pulls the water at a high rate of speed into the feed water nozzle and this, together with the force of the steam striking the water, drives the latter through the end of the feed water nozzle in a jet and into the valve chamber and since this is an inverted nozzle the steam expands.

Now when steam expands it loses in speed and, consequently, it gains in pressure until, at the lower end of the valve chamber, the pressure is so high it can pass into the boiler through the check valve, which opens only in the direction the stream is going.

The purpose of the overflow pipe is to allow the injector to get started and the water and steam escape through it.

The Steam Gauge.—It is a hard job to make a small steam gauge that is accurate and though

Fittings for Model Engines 131

I shall tell you how to make one and *calibrate* it, that is to mark the scale so that it will show the steam pressure of the boiler in pounds, my advice is to buy one of some reliable maker.

This kind of a steam gauge is known as a *Bourdon gauge* for the reason that it was first made by Bourdon, an instrument maker of Paris. First form a ring of $\frac{1}{32}$-inch thick sheet brass, $\frac{1}{2}$ an inch wide and $\frac{3}{4}$ inch in diameter; make a *butt joint* and solder it and drill a $\frac{1}{16}$-inch hole through the ring close to the joint.

Cut out two disks of sheet brass $\frac{1}{32}$ inch thick and 1 inch in diameter for the front and back of the gauge. Drill three holes through them with their centers $\frac{1}{4}$ inch from the edges and at equal distances apart; in the one to be used for the back drill two more holes in a line from the center and have each one $\frac{1}{4}$ inch on each side of the center; drill a $\frac{1}{16}$-inch hole in the center of the other disk which is the front one.

Make a pattern for the *spring box,* that is the hollow base which contains the pressure plate and the coiled spring, and have it cast. The pattern should be made in two pieces, curve the top part to fit the ring and hollow out the under part of it. Also hollow out the upper part of the bottom of the box and you will observe that both the top and the bottom have lugs on them.

When you get the castings, file them up and drill a 3/32-inch hole through the curved ends of the top; then drill a hole in each of the lugs and thread it for a screw and drill a 1/16-inch hole through the center of the box for the pressure rod to pass through.

Drill a hole in each of the lugs of the bottom part of the box, but do not thread them, and, finally, drill a 3/16-inch hole through the center of it and thread it to fit a 1/4-inch pipe.

Now for the works: make an angle support of a strip of brass 1/32 inch thick, 1/4 inch wide and 1 5/8 inches long; bend the edges as shown at B in Fig. 52 so that it stands 1/4 inch high; drill a hole in each end so that it can be screwed to the back; drill a hole in the center for the pin, or *arbor* as it is called, on which the pinion, that is a little cogwheel, is fixed inside the case and to the end of which the needle or pointer is screwed outside the case. Drill another hole 5/32 inch below the center and 3/32 inch to the left of it for the pin on which the toothed wheel turns;

and, lastly, drill a hole at the top and to the left, put in a pin and fix a flat spring to it.

Get a piece of *pinion wire*,[1] that is a ribbed wire used by clockmakers for making small pinions, or toothed wheels of any length, and cut it off ⅛ inch long, or you can use a small pinion

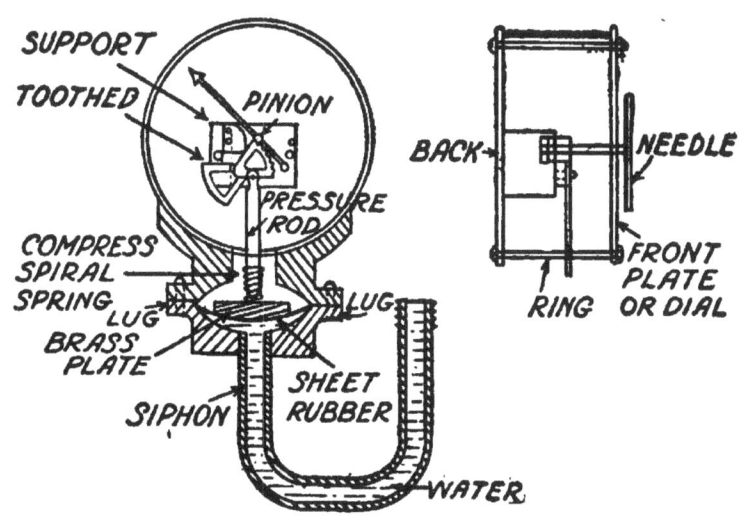

Fig. 52. How a Steam Gauge is Made

taken from an old clock. Set it on a steel wire ¾₁₆ inch thick and ⅞₁₆ inch long, thread both ends of the wire, or arbor, push the short end through the support and screw on a nut; and see to it that it runs very true.

Take another clock wheel, or buy a new toothed

[1] Can be bought of dealers in model makers' supplies.

wheel ⁵⁄₁₆ inch in diameter, drill a hole in two of the spokes ⅛ inch from the center and fix a pin in each one of them. To make the pressure rod sharpen one end of a piece of ¹⁄₁₆-inch wire ⅝ inch long; flatten the other end and drill a hole through it, and also flatten it at a point ³⁄₁₆ inch above the pointed end, drill it and fix a bit of wire in it for a stop for the spring. Now pivot the connecting rod to the toothed wheel and slip the rod through the hole in the bottom of the ring; then pivot the wheel to the support and screw the curved top of the spring box to the ring.

Next cut a piece of brass ¹⁄₁₆ inch thick, ¼ inch wide and ⁵⁄₁₆ inch long; round off one side of it with a file and make a dent in the center of the flat side of it with a center punch. Cut off a piece of thin rubber ½ inch wide and ¾ inch long; lay it on top of the lower part of the spring box and set the rounded brass plate in the middle of it with the flat side up.

Slip an open spiral spring over the pointed end of the rod, fit the bottom part of the spring box to the top, being sure that the pointed end of the pressure rod sets in the dent in the brass plate and screw the top and bottom of the spring box together tight.

Set the ring on the back plate; put the front plate on the ring with the arbor sticking through the hole and then bolt the two plates together. Make an index needle, or hand, of a bit of wire ⅝ inch long, flatten it, drill a hole 3/16 inch from one end and fix it on the end of the arbor. A minute hand from a watch makes a good needle.

Thread the ends of a piece of brass pipe ⅛ inch in diameter and 2¾ inches long; cork up one end and fill it with melted lead; when it is cold, bend the pipe into the shape of a U but with one end longer than the other end by ¾ inch; heat the bent pipe, or *siphon* as it is now called, until the lead melts and let it run out. Now screw the long end of the siphon into the boiler above the water line, of course; fill the siphon with water and screw the gauge to it.

Calibrating the Dial.—To calibrate the gauge you have made, that is to graduate the face of it, or *dial* as it is called, you must borrow an accurate gauge and fix it in the boiler together with your own.

As the pressure rises in the boiler you simply

mark your dial according to what the needle shows when it is compared to the first-rate gauge and that is all there is to it.

The Water Gauge.—To keep the boiler from

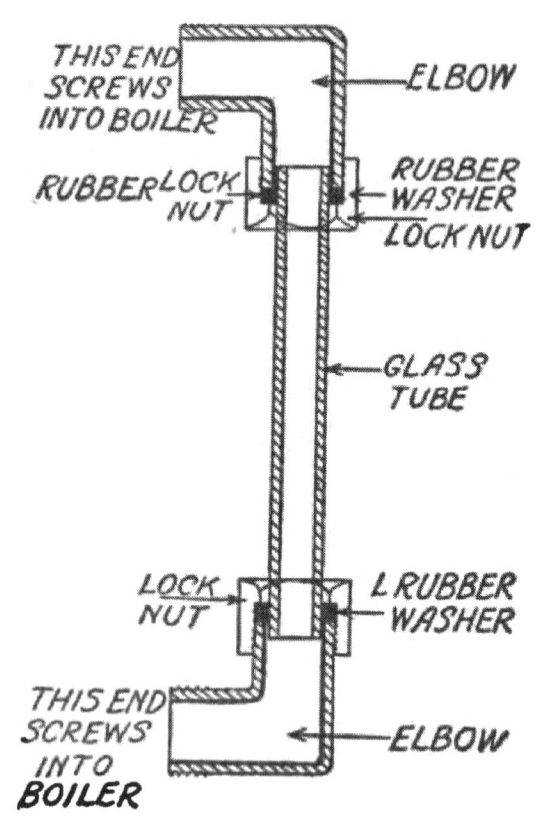

Fig. 53. Cross Section of a Water Gauge

blowing up for the want of water, you must fit it with a water gauge. This is a piece of glass tubing set into two elbows, one of which is

Fittings for Model Engines

screwed into the boiler below the lowest level the water ought to be allowed to get, and the other one is screwed into the boiler above the highest level at which the water ought to stand.

To make a water gauge for the boiler take two $5/16$-inch elbows and fit each one with a lock nut as shown in Fig. 53. Get a length of thick glass tubing for the water glass with an outside diameter just large enough to fit the inside of the lock nuts and elbows.

Next fit a thick rubber washer into each lock nut and slip the lock nuts on the tube. Put the ends of the tubes into the top and bottom elbows and then screw the nuts on them tight. The pressure of the lock nuts on the rubber washers will expand them, that is squeeze them out, and press them against the tube hard enough to make them steam-and-water-tight; at the same time it is easy to replace the glass tube should it get broken.

CHAPTER VII

A MODEL ATLANTIC TYPE LOCOMOTIVE

The Parts of a Locomotive: The Boiler: The Shell, The Smoke-stack, Bell, Sand Box and Steam Dome, The Saddle, Pedestal and Hanger, The Front Tube Sheet, The Steam Pipe and Throttle Valve, The Back Tube Sheet, The Crown Sheet, The Boiler Tubes, The Fittings—Making a Cardboard Model.

An *Atlantic*, or 4-4-2, type of locomotive as it is called by railroad boys, is one that has four small, or *truck wheels* in front, four *driving wheels* and two *trailing wheels*. It is a type of locomotive much used in the East for hauling fast passenger trains.

To make a model Atlantic type of locomotive that will look like a real one will not only take a lot of your time, but you will need a deal of patience, a fair skill in using tools and considerable cash to boot, but whatever time, trouble and expense you are put to you will feel repaid a hundred fold when you have it completed and in running order.

A Model Atlantic Type Locomotive 139

As in the case of the horizontal engine there are two courses open to you in building this locomotive and these are, (1) to make the patterns yourself and have them cast, and (2) to buy a complete set of castings and finish them up. My advice is to build the locomotive from the ground up, because when it is all done, it is really *yours;* on the other hand it will be much easier to buy a set of castings as each piece is cast exactly to size and shape.

To picture and describe each part in detail as I did the horizontal engine would take a whole book the size of this one, but after you have made the horizontal engine and the copper boiler which I told you about in the chapters that have gone before, you will have small trouble in building this model locomotive.

So that everything may be clear to you I have made *scale drawings* of the boiler and the engine and these will enable you to work intelligently, especially if you will follow the directions to the letter. One more thing, in reading the drawings just remember that a dotted line usually represents a part which is back of some other part and hence cannot really be seen. I have also used dotted lines to indicate the limits of measurements and sometimes to show where a piece of metal is to be bent.[1]

The Parts of a Locomotive.—Every locomotive, however large or small, is made up of the following main parts:

(1) The *boiler*

(2) The *engine*

(3) The *pilot,* or cowcatcher and

(4) The *cab*

The *tender* is an entirely separate, though none the less necessary, adjunct for it has a coal bunker and a water tank which supplies fuel and water to the locomotive.

The Boiler.—This boiler is of the *horizontal tubular* type and, like every other boiler, it consists of (a) the *boiler* proper, (b) the *smokebox* on the front end and (c) the *fire-box* on the rear end.

A modern locomotive boiler is made a little differently from the ordinary horizontal stationary boiler, and this makes it necessary to put the parts together in the order I have given.

---

[1] In engineering drawings only those lines that represent something that cannot be seen are dotted and all other lines are solid.

A Model Atlantic Type Locomotive 141

The Shell.—Get a seamless copper tube with a *wall*[1] ⅛ inch thick, 3½ inches in diameter and 16½ inches long; this forms the *shell,* as the outside of the boiler is called. Saw out one end of the tube, or shell, 1½ inches deep and 5¼ inches long as shown at A in Fig. 54. This done, drill three rows of holes around the shell so that the *sheets* can be riveted to it later. Drill the first

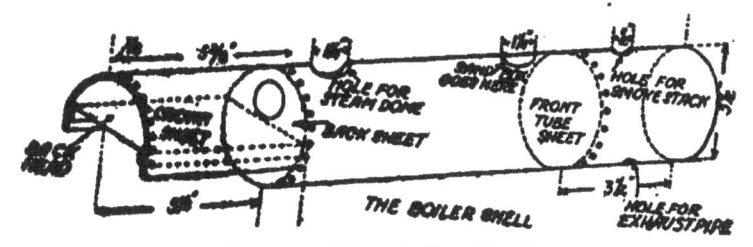

Fig. 54. The Boiler Shell

row of holes 3½ inches from the front end of the boiler, drill the second row 5⅜ inches from the back end and drill the last row ⅜ inch from the back end. Four holes only need be drilled ¼ inch from the front end as the smoke-box sheet is screwed to it instead of riveted.

Drill a hole for exhaust steam pipe and drill or cut out two holes in the top of the shell for the *smokestack* and the *steam dome;*

[1] The *wall* is the thickness of the metal of which the tube is made.

make the hole for the stack ½ inch in diameter and have its center 1⅞ inches from the front, and the one from the *steam dome* 1⅛ inches in diameter and have its center 6½ inches from the back end of the shell, all of which is shown in Figs. 54 and 55.

The Smokestack, Bell, Sand Box and Steam Dome.—The next thing to do is to make and rivet the *smokestack*, the *sand box*, the *bell frame* and the steam dome to the top of the shell as shown in Fig. 55. This is a *cross-section* view of the whole boiler and it will give you all the dimensions you need that are not shown in Fig. 54.

The easiest way to get a real looking smokestack, sand box and steam dome is to make patterns of them and have them cast in iron, then file them up and give them a coat of black enamel.

The smokestack should be ¾ inch in diameter at its base with a flange on it ⅜ inch wide, have a top 1 inch in diameter and be 1½ inches high. Make the sand box 1⅜ inches in diameter, with a ½-inch flange on the bottom, and 1½ inches high; bore a hole ½ inch in diameter in the top and fit a cover in it. A hole on each side ⅛ inch in

Fig. 55

Cross section of model Atlantic boiler and cross section of fire box and back of boiler

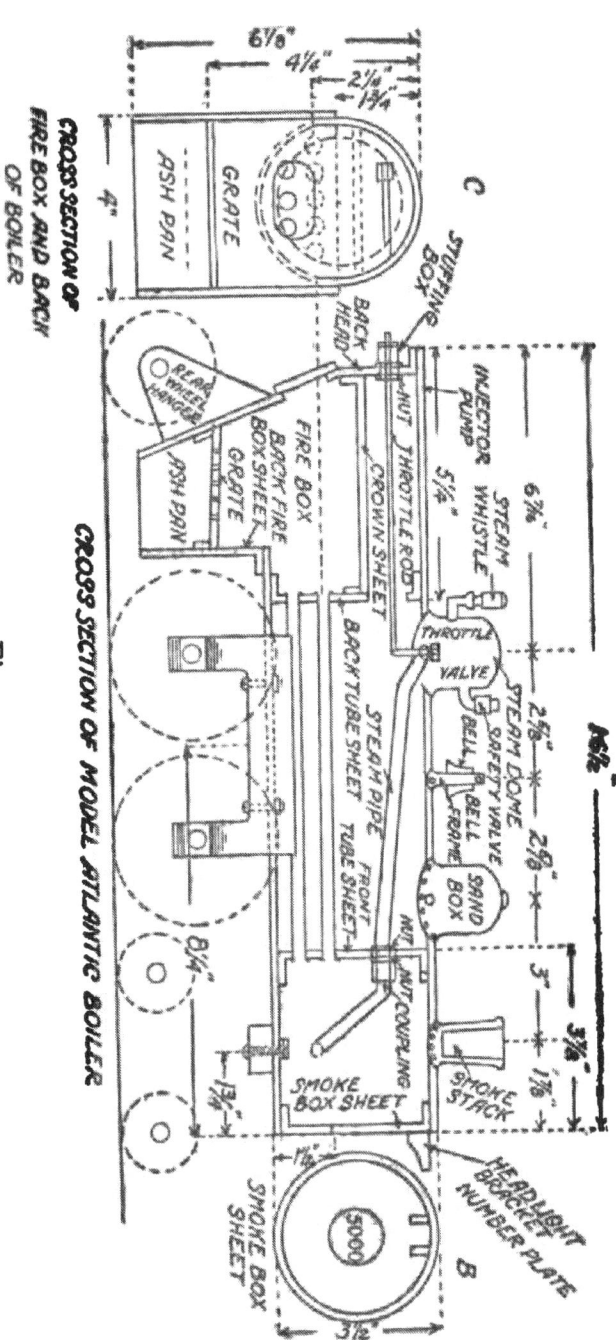

143

diameter near the bottom of the box can be drilled in the casting when you get it. These are for the *sand pipes*.

It's a hard job to make a pattern for the bell and have it cast, but you can buy a bell and make a frame for it. The steam dome is 1½ inches in diameter and 1½ inches high. Drill two ⅛-

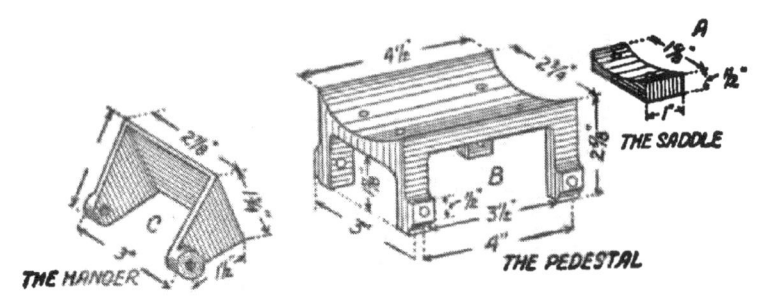

Fig. 56. The Saddle Pedestal and Hanger

inch holes on opposite sides of the dome as shown at A in Fig. 55; one of these is for the steam whistle and the other for the safety valve. The dome must, of course, have a wide flange so that it can be riveted securely to the boiler shell.

The Saddle, Pedestal and Hanger.—The next thing to do is to make the patterns for the *cylinder saddle, the driving wheel pedestal* and the *hanger* for the trailing wheels. These are shown at A, B and C in Fig. 56 and the dimensions are marked on them. Drill two holes through the saddle, four through the pedestal and two through the hanger so that they can be riveted to the boiler shell and fire-box as shown at A in Fig. 55.

Drill four holes through the hangers of the pedestal to form the bearings for the axle for the driving wheels and drill two holes in the ends of the hanger of the trailing wheels to form bearings for its axle.

When you have done all these things, smooth up the castings and then bolt the saddle to the under side of the boiler with its center 1¾ inches from the front end, see A, Fig. 55 again. Bolt the pedestal to the boiler in a line with the saddle and with its center 8¼ inches from the front end. You must screw these bolts up tight or else the boiler will leak.

The Front Tube Sheet.—Now make the *front tube sheet,* as shown at A in Fig. 57. Cut out a disk of sheet copper ⅛ inch thick and 4½ inches in diameter and hammer the edge of it over all round to form a flange ⅝ inch wide, when the sheet will just fit into the boiler.

This done, drill eight ¼-inch holes through the

lower part of it for the *boiler tubes*, though you can have fewer tubes if they are larger; drill another hole 1 5/16 inch in diameter through the

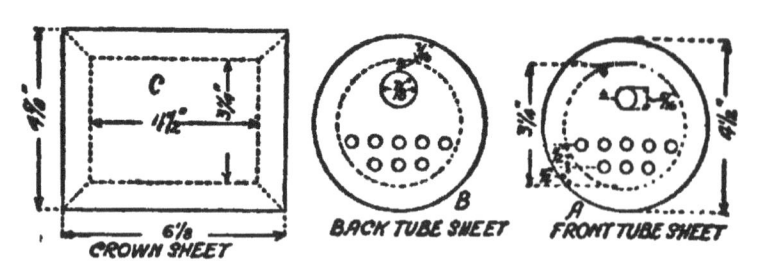

Fig. 57. The Front Tube, Back and Crown Sheets

upper part of the sheet for the steam pipe. Rivet the sheet to the shell 3 7/8 inches from the front end of the boiler, as shown in Fig. 55.

The Steam Pipe and Throttle Valve.—The *steam pipe*, which carries the steam from the steam dome, through the front tube sheet, then branches out into two pipes, and each one passes out of the smoke box on the side where it joins its respective cylinder; it is 5 1/16 inches in diameter and 7 1/2 inches long and threaded at both ends.

Bend the pipe as shown in Fig. 55 and screw a small stock-cock described in the previous chapter—on one end for the *throttle valve;* screw or otherwise fix a handle to the valve 3/4 inch long

A Model Atlantic Type Locomotive 147

and pivot a *throttle rod* $3/32$ inch in diameter and 7 inches long to the valve handle.

Screw a nut on the free end of the pipe, 1 inch down, push the end of the pipe through the hole in the front tube sheet and then screw another nut on the other side of the sheet; finally, screw a T on the end of the pipe. The throttle valve should now be in the center of the steam dome for it is here that the hottest steam generated by the boiler gathers.

The Back Tube Sheet.—Next cut out the *back tube* sheet as shown at B in Fig. 57 and make it the same size as the front tube sheet. Drill the same number of ¼-inch holes through it for the boiler tubes and also drill, or cut out, a ⅞-inch hole with its center $7/16$ inch from the flange. The purpose of this hole is to form a continuation of the boiler on back to the sheet; the throttle rod also passes through this hole; but it is part through before the back tube sheet is riveted in.

The Crown Sheet.—Cut out the *crown sheet*, see C in Fig. 57, that is the flat horizontal sheet that separates the upper back part of the boiler from the fire-box. Cut this sheet 4⅝ inches wide and 6⅛ inches long and hammer over the edge shown by the dotted line all around to make a flange ¾ inch wide when it will be the right size to fit; this is quite a particular job, because the corners must not be cut and they must be perfectly square.

Rivet one end of the crown sheet to the back tube sheet so that the upper surface of the former will be *flush,* that is even, with the lower edge of the large hole in the back tube sheet. Go ahead now and rivet the back tube sheet to the shell of the boiler at a distance of 5¼ inches from the extreme back end as shown at A in Fig. 55. Rivet the sides of the crown sheet to the edges of the shell on both sides as shown in Fig. 54.

The Boiler Tubes.—The boiler tubes, of which there are eight, are ¼ inch in diameter and 8 inches long. Set these in the holes in the front and back tube sheets and *expand* them and turn over their edges in exactly the same manner as in the vertical boiler described in Chapter V.

The sheet ought to come next but for this little boiler it, and the front sheet of the fire-box, can be made in one piece so we'll take up the making of the fire-box next.

The fire-box is made of four sheets riveted together. The back sheet is 5 inches wide and 4⅜ inches long. *Scribe,* that is scratch with a pair of *compasses,* a semi-circle having a *radius* [1] of 1 inch and another semi-circle, using the same center, the radius of which is 1¼ inches.

Divide the large semi-circle into 8 equal parts and then scribe a line from the middle of the semi-circle to each point which spaces it off. Cut out the smaller semi-circle and then cut the sheet through the scribed lines to the large semi-circle; hammer each piece, or *segment,* over until the end of the sheet fits around the boiler. The whole scheme is shown at A in Fig. 58.

Bend over each edge ¾ inch, making the width of the sheet 3½ inches as shown by the dotted line, and rivet a bar or strip of brass or iron ⅜₁₆ or ¼ inch thick, ½ inch wide and about 3 inches long across the sheet at a distance of 1⅛ inches from the bottom; this forms one of the *grate bar rests.* Now drill a hole in each segment of the sheet and a corresponding hole in the back lower end of the boiler shell and rivet them together.

For the front of the fire-box sheet use a sheet

---

[1] The radius of a circle is a straight line from the center of a circle to its circumference; hence the radius is half the diameter of a circle.

of copper 5 inches wide and 7½ inches long, as shown at B in Fig. 58. Scribe a semi-circle at one end in the middle of the sheet the radius of which is 1⅝ inches; then scribe a semi-circle whose radius is 2¼ inches, using the same center as before; cut it to shape and hammer over the rounded edge to form a flange ¾ inch wide. Cut off each side below the middle of the semi-circle so that the width of the sheet is 3½ inches, which is the width of the boiler.

The Fittings.—Now for the *fittings,* that is the steam gauge, the water gauge, and stuffing box for the throttle rod. Before the pipes for these fittings are fixed to the front fire-box sheet rivet on a *grate bar rest* like the one on the back fire-box sheet, 1¾ inches from the bottom as shown by the dotted line at B, in Fig. 58.

For the *door* cut out an oval hole 1 inch wide and 2 inches long in the sheet and ½ inch down from the *major chord,* as the straight edge of the semi-circle is called, and make a door ¼ inch larger all round, hinge it and rivet the hinges to the sheet so that the door will swing over the opening.

For the *water* gauge drill two ⅛-inch holes 1

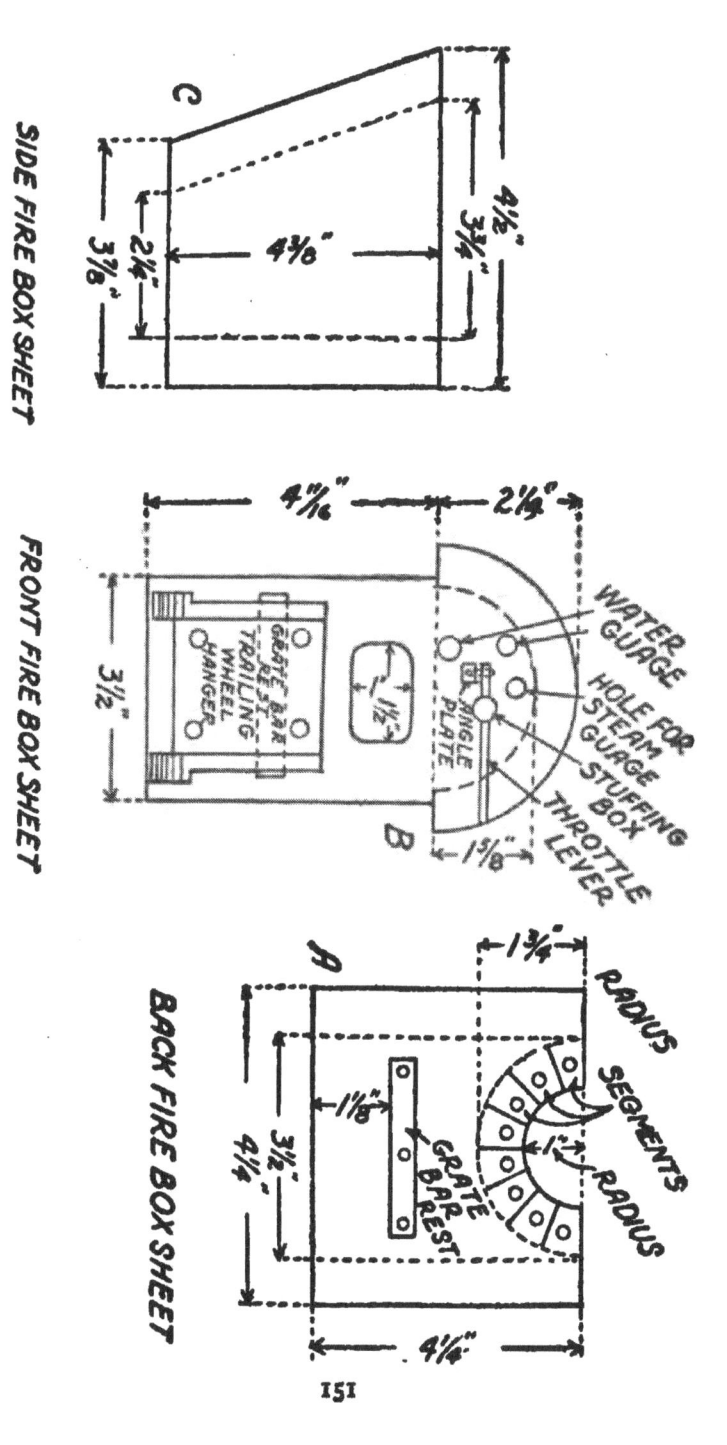

Fig. 58. The Side, Front, and Back Fire Box Sheets

inch apart from their centers on the left hand side and have the lower one as close to the *crown sheet* as possible. Put a piece of threaded pipe, ¾ inch long, through each hole and screw a nut on both ends of each pipe tight against the sheet.

For the *steam gauge* drill one more ⅛-inch hole in the middle near the top of the sheet and put in a threaded bent pipe 1 inch long and screw nuts on both sides as before.

For the *stuffing box,* through which the throttle rod passes, drill a 5⁄16-inch hole in the sheet ½ inch from the middle of the rounded edge. Take a piece of pipe 5⁄16 inch in diameter and ¾ inch long and thread it the entire length; screw a nut on one end down ⅝ inch, slip the end of the pipe through the hole and screw on another nut. Drill a 1⁄16-inch hole in a cap that will fit the pipe and your stuffing box is done. The position of all of these holes is shown at B in Fig. 58.

Make an *angle plate,* to pivot the throttle lever to, of a strip of brass 1⁄16 inch thick, ½ inch wide and ¾ inch long; bend over one end ¼ inch and drill a 3⁄32-inch hole in it and a ¼-inch hole in the other end; rivet it to the fire-box sheet in a line with and ½ inch to the left of the stuffing box.

A Model Atlantic Type Locomotive 153

The *throttle lever* can be made of a strip of brass $1/16$ inch thick, $3/16$ inch wide and 3 inches long; file one end to make a handle and drill a $1/8$-inch hole in the other end; $5/8$ inch from this hole drill a $3/32$-inch hole; pivot the end of the throttle lever to the angle plate.

Finally, rivet the *hanger* for the trailing wheels to the firebox sheet so that the center of the hole through which the axle passes is exactly 1 inch from the rim of the wheel which rests on the rail.

Now slip the end of the throttle rod through the stuffing box and rivet the semi-circular flange of the front fire-box sheet to the back edge of the boiler shell and then rivet the back flange of the crown sheet to the front fire-box sheet.

This done, bend in the front fire-box sheet at the line where the crown sheet is riveted to it, until its lower end is $1\frac{3}{4}$ inches from the back fire-box sheet; then pivot the throttle lever to the throttle rod.

For the sides of the fire-box cut out two sheets and make the tops of each one $4\frac{1}{2}$ inches wide and the bottoms $3\frac{7}{8}$ inches wide and have them $4\frac{5}{8}$ inches high as shown at C in Fig. 58; bend over the slanting side of each piece ¾ inch and then rivet the lapped seams together.

Now go back to the front end of the boiler and make the smoke-box sheet which is shown at B in Fig. 55. This is a copper sheet cut to the same size as the tube sheets but has a ½-inch flange. Solder or otherwise fix, a number plate 1 inch in diameter on the center of it so that it will project out ½ an inch.

Make a pair of brackets for the headlight and solder them to the smoke-box as shown at B in Fig. 55. You don't need to screw the sheet in the end of the smoke-box because the branched steam pipes have to be put in and fastened to the T. The headlight and the cab can be made after the boiler is mounted on the running gear and all the other finishing touches can be put on then.

Making a Cardboard Model.—Here is a bit of advice which, if you will take it, will save you a lot of trouble to say nothing about material.

Start at the beginning of this chapter and make a heavy cardboard tube following the dimensions given in the text and in the drawings. Mark and cut out each separate sheet and piece, bend them to shape and glue or paste them together until you have the whole boiler built up full size.

By making the boiler of cardboard first, you will see exactly where and how each seam is lapped and you will then have a working knowledge of how the boiler is constructed and this will prevent you from making a lot of mistakes when you come to making it up of metal. And now for the engine.

CHAPTER VIII

A MODEL ATLANTIC TYPE LOCOMOTIVE
(Continued)

The Parts of the Engine: The Cylinders, Steam Chests and Crosshead Guides; The Engine Truck Frame; Setting the Frame on the Truck Wheels; The Driving Wheels; The Side, or Coupling Rods; The Connecting Rods; The Link Valve Gear; The Trailing Wheels; The Pilot, or Cowcatcher; The Headlight; The Cab; Other Things to Do; The Tender—Finishing Up the Locomotive and Tender—How the Locomotive Works: How the Link Valve Gear Works.

Having completed the boiler the next thing to do is to build the engine and the running gear.

There is no special instruction you need to finish up the castings should you buy them ready-made but if you make the patterns and have them cast it is quite another matter.

The Parts of the Engine.—The engine of a locomotive is really formed of two separate and distinct horizontal engines that set on opposite sides of the smoke box.

Each engine consists of:

(1) A cylinder and steam chest with a cross-

head and crosshead block, a piston and piston rod, and a slide valve and valve-stem; these are mounted on,

(2) The *engine truck frame* which, in turn, is fixed to the boiler by means of the saddle;

(3) Two pairs of *driving wheels* with,

(4) Two *connecting rods* which connect the driving wheels with their respective piston rods;

(5) Two *side rods* to couple the driving wheels together, and

(6) A Stephenson *link valve gear* for reversing the direction of the engine.

The Cylinders, Steam Chests and Cross-head Guides.—The cylinders and steam chests with their pistons and slide valves are constructed like the horizontal engine described in Chapter IV, but the cylinders are turned around so that the steam chests are on top instead of on the side.

This brings the eccentric rod out of line with the slide valve stem; moreover, there are two eccentrics connected to each valve rod through what is called a *link;* the purpose of this arrangement is to enable the engine driver to reverse the direction of the locomotive at will.

Further, the cross-head guide for each piston

158 *The Boys' Book of Engine-Building*

must be mounted on the end of the cylinder instead of separate from it and, finally the cylin-

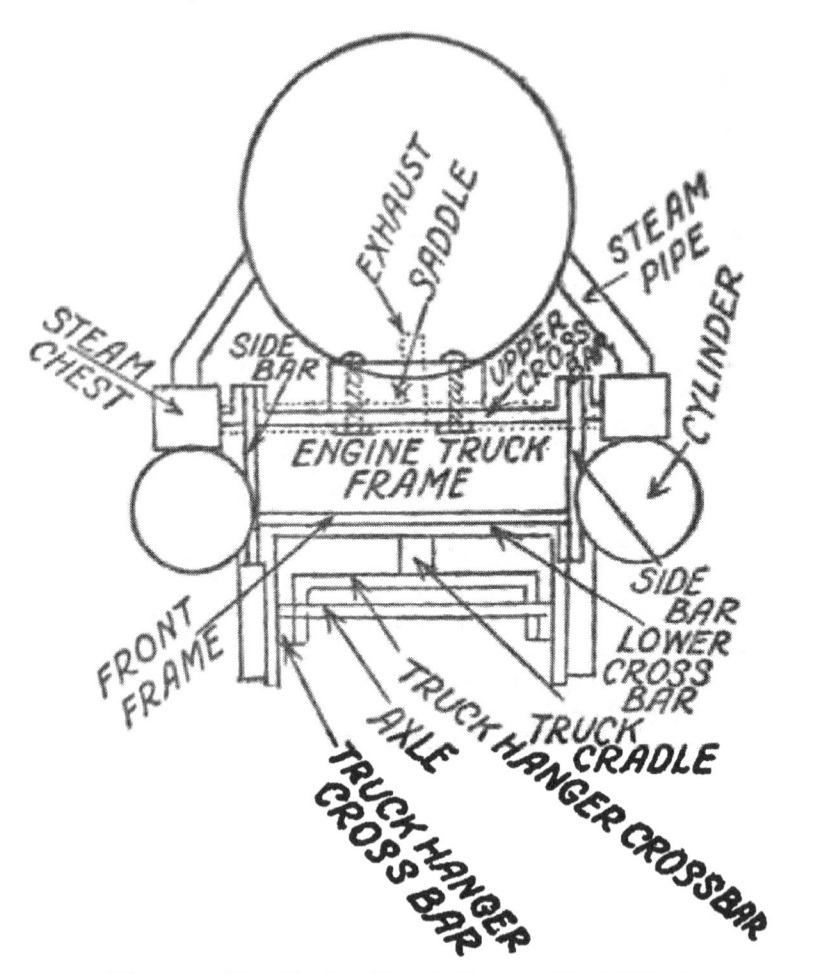

Fig. 59. The Engine Truck Frame (End View)

ders and steam chests are bolted to the sides of the *truck frame* as shown in the cross-section end view Fig. 59; consequently small changes must

A Model Atlantic Type Locomotive

be made in the design and construction of the engine and the cylinders and the steam chests must be made a little smaller than the horizontal engine previously described.

Make each cylinder 1 inch in diameter and 1¾ inches long; the steam chests should be ¾

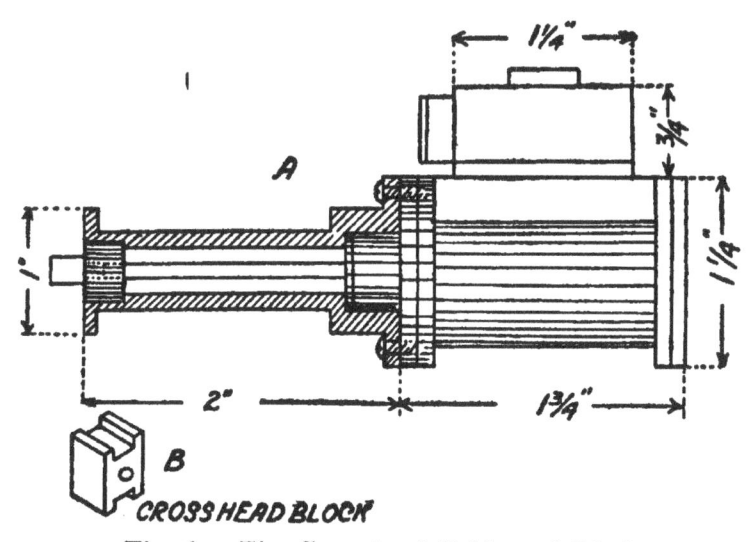

Fig. 60. The Cross-head Guide and Block

inch wide, ¾ inch high and 1¼ inch long, and make the piston rods and slide valves to fit as shown at A in Fig. 60; the chief sizes we marked on the drawings.

The crosshead guide is also shown at A in Fig. 60, as well as the way it is screwed to the cylinder head. The crosshead block that fits between the guides, and to which the piston rod is fixed and the connecting rod is pivoted as shown at B in Fig. 60.

The feet on one of the cylinders and steam chests must be reversed before the pattern is cast so that they will face each other as shown in Fig. 59.

The Engine Truck Frame.—After you have the cylinders and steam chests with their fittings made and in working order you can begin work on the *engine truck,* that is the part which supports the cylinders and steam chests and on which the smoke-box of the boiler sets on the front, or *truck wheels,* as they are called.

Make the *truck frame* first and build it up of sheet brass 3/32 or 1/8 inch thick, make two cross bars each 1 inch wide and 3 3/8 inches long, bend up each end 1/2 inch and drill a 3/32-inch hole through it. Drill two holes in the upper bar so that the saddle can be bolted to it and to the smoke box as shown in Fig. 59, though this is not done until the frame is finished.

Drill a 1/8-inch hole through the middle of the lower cross bar for the *engine truck cradle* as it is called. Make two *side bars,* see Figs. 59, and

A Model Atlantic Type Locomotive   161

61 and have them 1¼ inches wide and 2 inches long, and drill a hole in each corner, so that you can bolt the cross-bars, the side bars and the cylinder and steam chest supports together as shown in Fig. 59.

Before bolting them together cut out a sheet of brass ⅛ inch thick, 1¾ inches wide and 3½ inches long and bend over one end ½ inch. Drill four 3/32-inch holes in this plate at one end ⅝ inch apart and two holes through the angle plate ¾ inch apart as shown in Fig. 61. Screw or rivet this plate to the lower cross bar and you have the *front frame* ready for the *buffer beam*.

The *truck hanger*, which supports the engine truck frame and forms the bearings for the truck wheels, is made of two brass bars 3/16 inch thick, ⅝ inch wide and 2½ inches long; the bars can be made of a strip of brass or, better, they can be cast in brass. Whichever kind you use the ends must be set at an angle of 90 degrees to the cross bar to form bearings for the wheel axles.

Drill a ⅛-inch hole through the end of each bar and two 3/32-inch holes, ⅜ inch from the middle of each bar. Screw the truck hanger side bars to the truck hanger cross bar and then pivot the lower cross bar of the frame to the truck hanger cross bar; to do this make a collar,

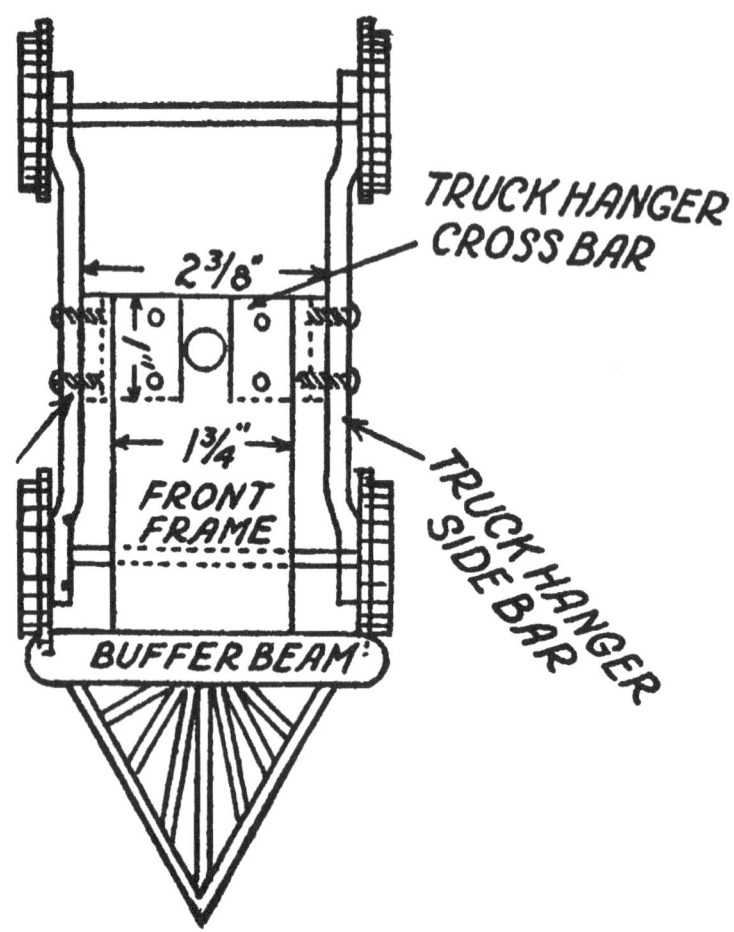

Fig 61. Top View of the Truck Frame

or, as engine men call it, a *truck cradle*, ⅝ inch in diameter and ⅝ inch high and drill a ⅛ inch hole through it lengthwise. Bolt the lower cross bar to the truck hanger cross bar with the cradle in between them as shown in Fig. 59.

Setting the Frame on the Truck Wheels.—The standard gauge of a railroad track in the United States is 4 feet 8½ inches but for your model locomotive a track gauge of 3³⁄₁₆ inches is the right width. The only outstanding thing to do now as far as the engine truck frame is concerned is to put on four truck, that is, front wheels.

These wheels must be regular car wheels, that is flanged, as shown in Fig. 61, have a face ⁵⁄₁₆ inch wide and a diameter of 1½ inches. Iron castings of car wheels of this size can be bought for 5 cents each and brass castings of the same size costs 30 cents each.

Drill a ⅛-inch hole through the center of each wheel and *tap it,* that is cut threads on it; cut off two axles of ⅛-inch soft steel rod and make each one 3¾ inches long; thread the ends, screw a wheel on one end of each one with the flange on the inside. Next, slip the axle through the holes in the truck hanger bar and screw on the other wheel; then do the same thing with the other axle and wheels.

Raise up the front end of the boiler and set

it on the saddle; set the saddle on the upper cross bar, and this completes the engine truck frame.

The Driving Wheels.—These should be 3½ inches in diameter with a 5/16-inch face and each one must be spoked and have a *counterbalance weight* cast on one side to balance the weight of the connecting rod and side rod opposite it. A cast iron driving wheel in the rough, but spoked and counterbalanced, can be bought for 25 cents each, while brass ones cost 50 cents each and you will need two pairs. Drill a 3/16-inch hole through the center of each one and thread it to fit a ¼-inch threaded axle.

Drill and thread each wheel to fit a threaded ⅛-inch *crank pin* opposite the counterbalance weight; the bearings of the connecting rods fit on the crank pins on the rear wheels, and the bearings of the side rods fit on the crank pins on both the front and back wheels as shown in Fig. 62.

Cut off two pieces of soft steel rod ⅛ inch in diameter and ¾ inch long and thread the ends. Screw one of these pins into each of the front wheels; cut off two more pieces of the ⅛-inch rod and have these 1¼ inches long and threaded at both ends. Screw one end of each of these pins, which form the crank pins, into the hole opposite the counterbalanced weight.

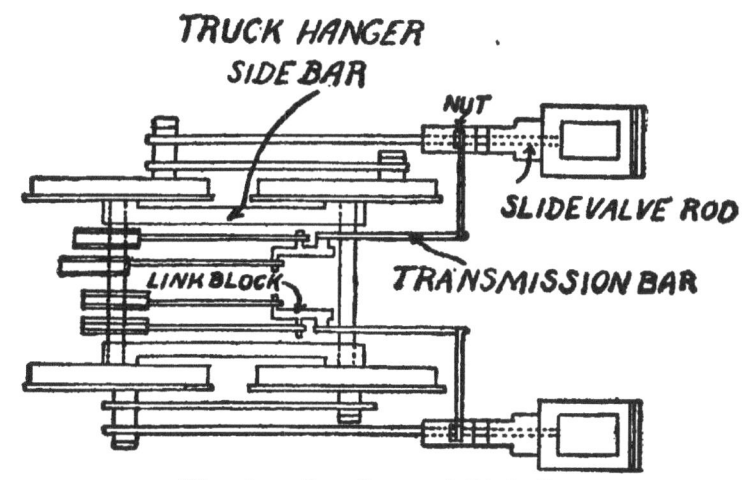

Fig. 62. Coupling and Link Gear

The Side, or Coupling Rods.—You can either make patterns of and have the side rods cast, thereby making them look like those on a regular locomotive, or you can make them out of a strip of brass ⅛ inch thick, ⁵⁄₁₆ inch wide and have each of them 3¾ inches long.

Drill a hole in each end of each side rod ⅛ inch in diameter and exactly 2 inches apart from their centers and have the holes fit the crank pins to a nicety.

The Connecting Rods.—Make two connecting rods in the same way that you make the side rods and have them of the same thickness and width, but 6½ inches long; drill a hole in one end of each one; the other end is to be fitted with a joint, or fork, like the horizontal engine so that it can be pivoted to the cross head block with a pin.

Before fitting the bearings of the rods to the crank pins slip a small washer over each pin in each one of the four driving wheels; this done, push the crank pins through the holes in the side rods, and on the ends of the crank pins on the front driving wheels screw on two nuts; the inside nut must not be put on too tight or the pin will not turn easily in the bearing of the side rod. The outside nut must be screwed on tightly to keep the inside nut from working loose, or to *lock* it as it is called.

Slip a collar, ⅜ inch high, over each crank pin on each of the rear driving wheels, then put on the connecting rod and screw on a couple of nuts as before.

Screw one end of both the front and rear axles to a pair of the driving wheels and push the free

A Model Atlantic Type Locomotive 167

ends of the axles through the bearings on one side of the pedestal as shown in Fig. 63; before screwing the driving wheels on the other

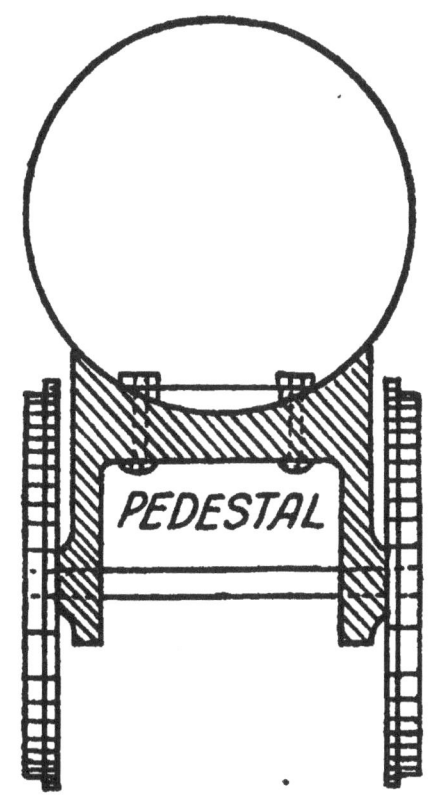

Fig. 63. How the Driving Wheels Are Mounted

side of the pedestal to the axles the *link valve gear* must be made.

The Link Valve Gear.—The purpose of this *gear* is to reverse the engine. It must be made

168 *The Boys' Book of Engine-Building*

and the eccentrics, which are connected to it, must be *keyed* or otherwise fixed to the axle of the rear driving wheels before the axle is pushed on through the farthest bearing of the pedestal.

The first thing to do is to know how the link valve gear is made. A side view of it, showing

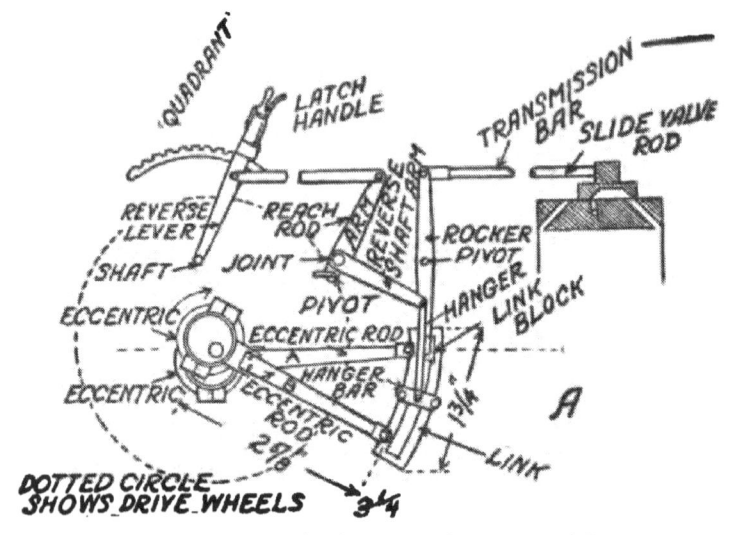

Fig. 64. Stephenson Link Reversing Gear (Forward)

the position of the *reverse lever*, link and eccentrics when the gear has been set to go ahead is given at A in Fig. 64. The position of the reverse lever and link when the engine has been reversed to back up is shown at B.

Now to make the engine run smooth and even the pistons are set 90 degrees apart, which means

A Model Atlantic Type Locomotive 169

that while one piston is in one end of the cylinder the other piston is in the middle of its cylinder and this brings one of the crank pins to the extreme front or back of the wheel while the other crank pin is at its highest or lowest point on its wheel, all of which is shown in Fig. 62.

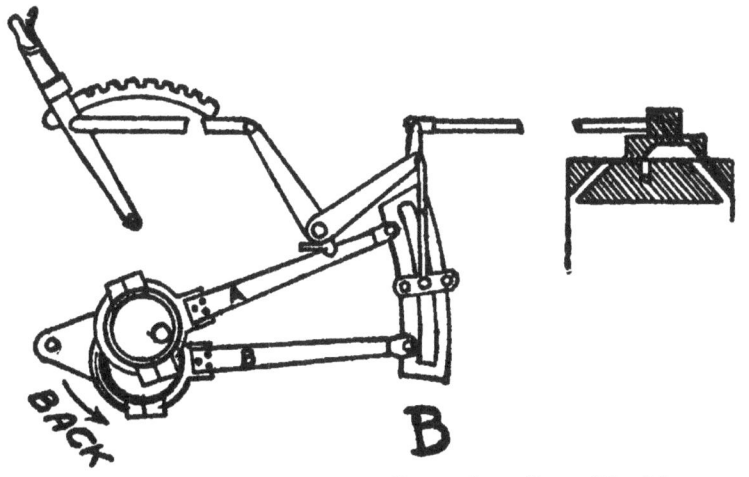

Fig. 64. Stephenson Link Reversing Gear (Back)

Not only this, but the eccentrics must also be set at 90 degrees from each other and of course each one of each pair—that is whichever one is working the slide valve—must be set 180 degrees, or opposite to its respective crank pin when it is brought into the operating position by the reverse lever.

To make the link valve gear is not so hard a

job, but to set it so that the eccentrics will work just right will give you all the trouble you are looking for.

Make four eccentrics, each of which is ⅝ inch in diameter and 2⅝ inches from the center to the end of the rod, and drill a hole in each one for the axle, ¼ inch in diameter and 3/16 inch out of the center as shown at A in Fig. 64; and this will give a *throw* of ⅜ inch.

Next, make the links, one for each pair of eccentrics; to get the proper curve the inside *arc* should be a part of a circle having a diameter of 4½ inches and each of the outer circles should be ⅛ inch larger than the next inside circle. If you will mark out a link the exact size and shape shown at A or B in Fig. 64, of the size given above, and saw it out of a smooth strip of wood, ⅛ inch thick, with a fret saw, have it cast in brass and then file it up, you will have a link that will work better than one made of sheet metal.

The *bosses,* as the projections on the inside of the link near the ends are called, are what the ends of the eccentric rods are pivoted. The bosses in the middle of the link inside and outside are what the *hanger bars* are fixed to.

These must project out and away from the curved slot in the link far enough to let the *link block* slide back and forth easily; this block is made in exactly the same fashion as the cross head block shown at B in Fig. 59.

Now make the *rocker* of a strip of brass ⅛ inch thick, ¼ inch wide and 2¼ inches long; drill a hole in each end and one in the middle; pivot one end to the link block on one side and the other end to the slide valve rod by a *transmission bar*, which is simply a continuation of the slide valve stem. The middle part of the rocker is pivoted to the side of the pedestal.

This done, make the hanger of a ⅛-inch thick strip of brass, ³⁄₁₆ inch wide and 1¼ inches long; drill a hole through each end and pivot one end to the hanger bar but on the other side of the link to which the rocker is pivoted. The *reverse shaft arm* is a piece of brass ⅛ inch thick, ¼ inch wide and 1⅜ inches long with a hole drilled through both ends; it is jointed to the *reach rod arm* which is of the same size except that it is 1½ inches long, drill three holes in it as shown at A in Fig. 64 and put a pin in the lower end so that it can be pivoted to the pedestal.

The *reach rod* is a long bar reaching from the upper end of the reverse shaft arm to the reverse lever in the cab. It can be made of iron or brass ⅛ inch thick, 3⁄16 inch wide and 4¾ inches long and is pivoted to the *reverse lever* at about the middle of the latter. The reverse lever should be 2⅜ inches long with a *latch handle* and its lower end is pivoted to one end of a shaft ¼ inch in diameter and 1 inch long, the other end of which is fixed to the side sheet of the fire-box.

The latch handle works in a *notched quadrant*, the ends of which are also fixed to the side of the fire-box.

Two complete sets of all these parts are needed, except the quadrant reverse lever and reach rod arm. The lower ends of both reach rod arms are fastened to the ends of a shaft which is held in place and yet free to turn in a fixed bearing. Now when the reverse lever is pushed ahead or pulled back, both link valve gears will be operated by it.

When you have the two complete sets of link gears made, slip the eccentrics over the rear axle, push the latter through the bearing in the pedestal and screw on the other driving wheel.

A Model Atlantic Type Locomotive 173

This done, go ahead and couple on the other side rod and the other connecting rod and, finally, pivot the connecting rods to the crosshead block with a pin.

Before making the eccentrics it is a good plan to make a large working model of the whole link gear out of thin wood and cardboard and see exactly how the *link motion* works and learn how it acts on the slide valve; then when you make the link gear out of metal you will know just what you are about.

The Trailing Wheels.—These wheels are 1¼ inches in diameter and all you have to do to put them on is to drill a 3/16-inch hole in the center of each wheel and thread it, then make an axle of ¼-inch rod and thread it at both ends. Screw one end of the axle to a wheel, push the axle through the bearings in the hanger as shown at A in Fig. 55, and screw on the other wheel.

The Pilot, or Cowcatcher.—While there are very few animals that are given the chance to horn an engine off of the track the cowcatcher is still the front and foremost part of a locomotive in this country.

Make a *buffer beam* of wood, ½ an inch square

and 3½ inches long, as shown in Fig. 61, and screw it to the angle plate which is mounted to the *front frame*. Build up the pilot of fifteen wood strips and glue them to a triangular wood frame so that a sharp point sticks out in front of the buffer beam 2⅛ inches; make it 3½ inches high and screw it to the buffer beam. The pilot can be braced to the buffer beam underneath and it can be braced to the sides of the smoke-box on top.

The Headlight.—While this can be a *dummy* cut out of wood, it is better to make it out of tin and solder the seams. Form a reflector of bright tin and put a 1 candle power *tungsten* electric lamp in front of it; run wires from it to the tender where you have a dry cell battery.[1] When the lamp is lit, the beam of reflected light will produce a pretty and realistic effect in a darkened room.

The Cab.—This can also be made of heavy sheet tin with the seams soldered together. Have it 4 inches long, 4½ inches high and 5½ inches

---

[1] Complete instructions for wiring up electric lamps will be found in "The Book of Electricity," by the present author and published by D. Appleton and Co.

wide, but let the curved roof project 2 inches, making the length of the cab 6 inches over all and put a ventilator in the top of it.

Other Things to Do.—There are a hundred and one other things to do if you intend to put on all the attachments that are to be found on a real locomotive. You should by all means put on a *hand-rail* and a *running board;* then there are the springs, if you want to go to the trouble of making them; the *air brakes,* which, if you made the *compressor* right and the *three-way cock* that controls the air in the *line pipe,* will take you half as long as it did to make the engines.

The Tender.—Every locomotive must have a tender to carry a supply of fuel and water. Make the tender of heavy sheet tin 5¼ inches wide, 3½ inches high and 9 inches long. The body of the tender is made double all round except in front so that water can fill the space between the sides and back walls or tank and coal can fill the middle and open space or bunker.

The tender has eight wheels and all of them are the same size as the truck wheels of the locomotive. The wheels are mounted in sets of two pairs to a frame and the body rests on the front and back frames.

A pair of *buffer castings* is used to couple the tender to the locomotive and a piece of sheet metal which is fixed to the floor of the cab rests on the floor of the tender so that the fireman won't fall through between them when he is shoveling coal from the coal bin into the fire-box.

Finishing Up the Locomotive and Tender.—Following the practice of real locomotive building, get a can of black enamel and paint the boiler from the front to the back tube sheet with it.

Also paint the cab, the wheels, the head-light and the pilot with it to make them a shiny black. The steam dome, sand box, smokestack, smokebox and that part of the furnace which shows below the cab should be painted a dull black and this can be done by simply adding turpentine to the enamel. The locomotive complete is shown in Fig. 64.

How the Locomotive Works.—Of course, when a fire is built in the fire-box the water is heated as in any other type of boiler; the hottest

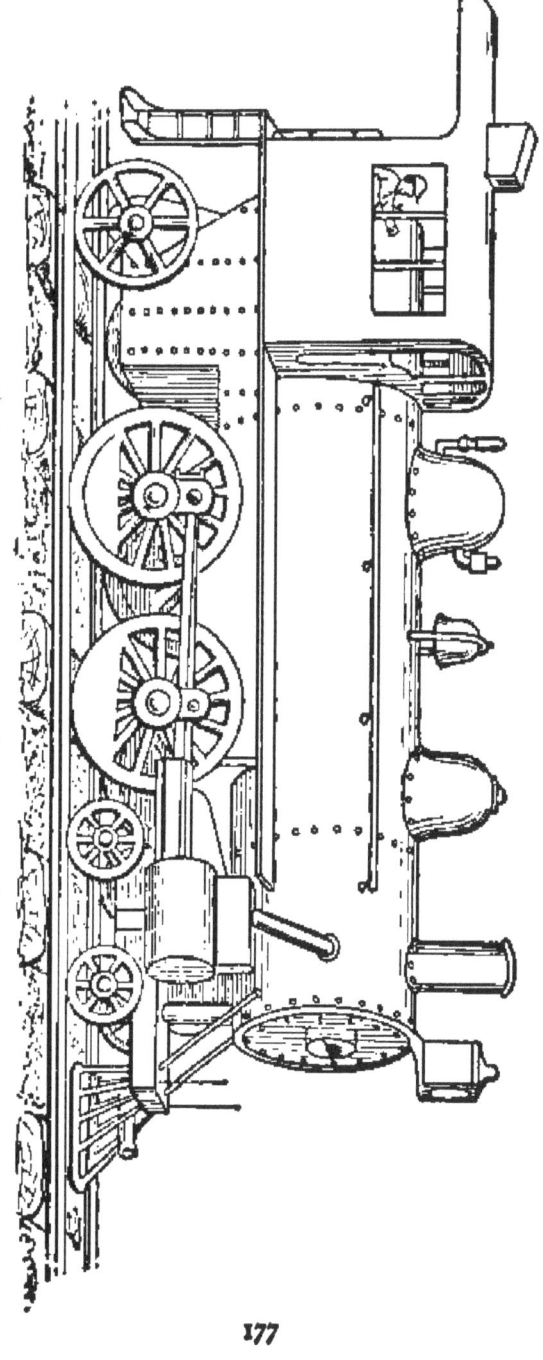

Fig. 65. Model 4-4-2 Atlantic Type Locomotive When Done

steam rises to the highest point which is the steam dome; from the dome it passes through the *throttle stand pipe,* as the bent up end of the main steam pipe is called, the amount going through depending on the steam pressure and the distance the throttle valve is opened.

Having reached the T in the smoke-box, the steam passes out through the two branch pipes and thence into the steam chests where it is distributed to the ends of the cylinders. Since there are two sets of steam chests and cylinders the power is equally divided and applied on both sides of the locomotive.

Again, since the pistons are set at 90 degrees apart, one of the pistons is forced forward half way while the other one is being forced back half way and then both pistons travel half the length of their strokes in the same direction; the result of these actions makes the engines run steadily.

As the used steam from the cylinders exhausts into the smoke-box it blows out through the smoke-stack; this sets up a draft through the fire tubes and fire-box and makes the fire burn better.

A Model Atlantic Type Locomotive 179

Now since each cylinder exhausts twice during every revolution of the driving wheels to which its piston is connected and since the piston on the other side is set so that it is midway in its cylinder while the other piston is at the end of its stroke there are four exhausts to every revolution of the driving wheels, as you have probably observed not once but many times, in real locomotives, especially when they are just starting. One thing more and that is the action of the link valve gear.

How the Link Valve Gear Works.—After you have made the link valve gear, or even a model of it, you will have a clear understanding of the how and the why of it.

The purpose of the link valve gear, as you know, is to reverse the engine and to do it the slide valve must be pulled or pushed over the inlet port of the cylinder so that the first flow of steam into it will force the piston in the direction needed to make the engine go ahead.

Suppose that the reverse lever, see Figs. 62 and 64, is in the middle of its quadrant, the link is then set in such a position that the link block at the lower end of the rocker is in the middle of the link and consequently half way between the pair of eccentric rods which control it.

When the link block is in this position the engine cannot work because the slide valve is brought to the middle of the steam chest and the action of the eccentric rods on it is so slight that it cannot move far enough in either direction to cover the inlet ports alternately.

But when you throw the reverse lever forward as far as it will go, it will push the link down until the block is at the top of it and this will throw the slide valve to the end of the steam chest so that the first flow of steam will be through the port which will force the piston in the direction to make the wheels revolve forward and thus carry the locomotive ahead.

When the reverse lever is thrown forward the eccentric rod A, which is the upper one, is in a line with the link block and works the slide valve through the rocker which is pivoted to the block; in this case the eccentric rod B drops down and though it moves to and fro it does not in any way affect the movement of the slide valve.

On the other hand, when the reverse lever is thrown back, it pulls the link up and this brings the eccentric rod B, which is the lower one, into line with the link block; when this takes place the eccentric rod A is raised out of the way and though it moves back and forth it does not in any way interfere with the motion of the slide valve set up by the eccentric rod B.

CHAPTER IX

STEAM, THE GIANT POWER

The Stuff that Steam is Made of—How Water is Formed—What Heat Does to Water—How Water is Made to Boil—Getting Up Steam—About Steam Pressure—How Steam Acts—Work and Horse Power—How to Calculate the H. P. of Your Engine—How to Calculate the Size of a Boiler—How to Calculate the Heating Surface of a Locomotive Boiler.

The Stuff that Steam is Made of.—You know, of course, that steam is formed of water and you also know that water is good to drink and to go swimming in, but do you know just what the stuff is really made of? If not read on. Water is made up of two gases, one of which is *hydrogen* and the other is *oxygen*. Now hydrogen is the lightest substance known, being 14½ times lighter than the air we breathe and 16 times lighter than the oxygen which keeps us alive.

Pure hydrogen burns with a very hot flame and is so nearly the color of daylight it can hardly be seen in it. Further, it gives off 5 times as much heat, when burned, as the same weight of coal.

Oxygen is even a more important gas than hydrogen and, fortunately, there is more of it in nature than any other kind. It is a great gas to combine with other chemical substances and when anything burns you may know that it is the chemical elements in it that are combining with the oxygen.

If you filled a quart measure with hydrogen and another quart measure with oxygen and weighed them, you would find that the oxygen weighs 16 times as much as the hydrogen.

Chemists have found that if 1 part of hydrogen by weight could be combined with 8 parts of oxygen, also by weight, together they would form water, but they also found that neither hydrogen nor any other substance will combine with less than 16 parts of oxygen by weight.

But it is an easy matter to combine 16 parts of oxygen with 2 parts of hydrogen, consequently you can write the *formula* for water thus: hydrogen 2 parts to oxygen 16 parts, or you can use, if you like, the letters H and O that stand for hydrogen and oxygen in chemistry, or *symbols* as they are called, where H stands for 1 part of hydrogen and O for 16 parts of oxygen and then write it H_2O.

How Water is Formed.—To simply put 2 parts by weight of hydrogen and 16 parts by weight of oxygen in a bottle and shake them up is not enough to make them form water; instead they will remain separate gases as before, but, being so close together, if you light them with a match, or by an electric spark, they will explode violently and water will result.

One way of making these gases combine chemically to form water is to pass hydrogen over a compound that has oxygen in it, such as *copper oxide,* and heating them. At ordinary temperatures hydrogen will not act on the copper oxide, but if you heat them in the flame of a Bunsen burner the hydrogen will combine with the oxygen to form water, and leave the copper behind.

What Heat Does to Water.—While it is quite easy to produce water from hydrogen and oxygen, as I have just described, it is not an easy matter to make the rule work the other way about, for to change it back again into its two constituent gases, or *decompose* it, as it is called, you must raise the temperature to at least 2000 degrees Fahrenheit.

You can, however, easily decompose water by *electrolysis,* that is by passing a current of electricity through it, provided you use the right kind of apparatus.[1]

When you *ignite,* that is light, an alcohol lamp or a Bunsen burner, or any kind of a fire, heat is developed and heat is a form of *energy*. Now heat is made up of little to and fro movements of the particles of gases that are burning or of the particles of matter of a substance that is warm or hot.

These *vibratory motions* of the *molecules* of *matter,* as they are called, of the flame or fire set the particles, or molecules of matter of which the boiler is formed, into rapid vibration, when the energy of the hot molecules is soon imparted to the molecules of water inside of it and these begin to vibrate, which warms the water first, then makes it hot and finally causes it to *boil.*

How Water is Made to Boil.—*Ebullition* and

[1] This can be bought of the L. E. Knott Apparatus Co., Boston, Mass.

boiling mean exactly the same thing only the first is the didactic, technical word and the latter is the simple, everyday word, so we'll use it instead.

When water begins to boil, the first thing that takes place is that the air, which is always plentifully mixed with it, is forced up from the bottom in little bubbles and these rise to the top and pass into the air without making any noise about it.

As the *temperature* of the water, that is the heat in it, is raised the molecules of water that rest on the bottom of the tea kettle, or the surface of the boiler, next to the fire, get so hot that steam bubbles are formed and the force of these is greater than the combined weight of the water on them and the pressure of the air above them can stand.

Getting Up Steam.—When water is heated to 212 degrees Fahrenheit it boils and no matter how much more heat is applied the water will not get any hotter.

This curious action is due to the fact that all of the heat that is used to change the water into steam is carried off by and in the steam, and this is made to do useful work in the steam engine as you will presently see.

Heat, like every other kind of energy, may be either *kinetic,* that is in motion, or *potential,* that is at rest; but wherever energy is at rest it is always ready to get into motion if it is given half a chance.

Now when water is boiled the heat tears the molecules from it and from each other and throws them into violent motion and the more it is heated the greater this molecular motion becomes and this is what is called the *sensible heat* of the steam.

But not all of the kinetic energy of the heat is used in this way for some of it is stored up in the steam and this *potential energy,* which is called *latent heat,* charges the steam with energy just like electricity charges a Leyden jar.

And don't forget that energy of motion can change into energy at rest, and that energy at rest can change into energy of motion, and, moreover, whether you are dealing with electricity or steam these changes are made forth and back with amazing freedom and in the twinkling of an eye.

About Steam Pressure.—While steam acts like a gas it is called a *vapor* because in physics the word *gas* is used to mean a gaseous form of matter which can only be made into a liquid by applying considerable cold and pressure to it.

On the other hand a vapor is a gaseous form of matter made from some liquid, or solid, substance by heating it and which, as soon as the vapor cools, will condense and return to its liquid, or solid, state again. But steam is a gas just the same and it consists of the two combined gases that were chemically united to form water.

Like hydrogen, or oxygen, or any other gas, the molecules of steam are elastic—far more elastic than solid rubber balls—and they are continually shooting in and out in every direction. The speed at which a hydrogen molecule travels is more than a mile a second and the speed of a molecule of steam is very nearly as swift.

Each and every molecule shoots in a straight line at the same high speed until it either hits another molecule, when both are stopped for an instant and their directions changed, or until it strikes the side of the vessel which holds them.

The bombardment of the molecules on the containing vessel, whether it is a teapot, a boiler, or the cylinder of an engine, is so fast and furious that the sum of all these little molecular blows result in a powerful force that is able to do work, and this is what we call *steam pressure*.

The pressure of steam can be measured either in *atmospheres* or in *pounds*. In England both scales of measurement are used, but in this country only the latter is employed.

The word *atmosphere,* of course, means the air we breathe, but it also is used to indicate a pressure of 15 pounds to the square inch, since the weight of a column of air 1 inch square measured from the upper limits of the atmosphere to the earth at sea level is 15 pounds. A scale of this kind is shown at A in Fig. 66. To read it you have to convert the steam pressure in atmospheres to pressure in pounds, that is 1 atmosphere means 15 pounds, 2 atmospheres 30 pounds and so on. A scale graduated so that the pressure can be read directly in pounds is shown at B.

How Steam Acts.—*In the Boiler.*—Since the smallest amount of steam will fill the largest boiler and keeping up the heat will increase the

190 *The Boys' Book of Engine-Building*

pressure of the steam you must never let the water in the boiler fall below a certain level or there will be an explosion.

Fig. 66 A. A Steam Gauge Scale in Atmospheres

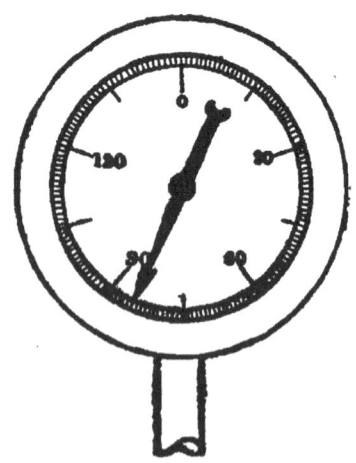

Fig. 66 B. A Steam Gauge Scale in Pounds

In the Engine.—When steam is admitted by the slide valve into the cylinder it is not only under the pressure of the steam developed by the sensible heat as it comes from the boiler, but after the steam in the cylinder is cut off by the slide valve its latent heat, which is energy at rest, begins to change into energy of motion, and this makes the steam keep on expanding and forcing the piston along.

To make steam do as much work as possible it should not be allowed to exhaust until its expansion has reduced its pressure almost to that of the pressure of the outside air, when it will escape silently; but when it puffs out with a noise you will know that some of the energy of it is being wasted.

It takes a *high pressure* engine to exhaust directly into the open air, because the pressure of the steam in the boiler must be more than 15 pounds to the square inch, that is, it must be higher than the pressure of the outside air. Nearly all small engines and locomotives are of the high pressure kind.

It is more economical, though, to connect the exhaust of the engine with a *vacuum chamber*, that is, an airtight vessel in which the pressure of the air is less than 15 pounds to the square inch.

This is done by making the steam exhaust into the chamber and *condensing* it there by admitting a jet of cold water; the condensed steam and water and any air that may have leaked into the chamber are then pumped out with a small pump.

Engines fitted with vacuum chambers are called *condensing engines* and nearly all the engines of the day of Newcomen and Watt were of this kind.

Work and Horse-power.—Now that you have seen how steam is charged with the energy of heat and how it has the power to do work, the next thing is to know how this energy can be measured.

The most convenient way and the one generally used is to simply change the heat energy into mechanical energy which is easier to measure. In the steam engine the heat energy of the steam acts on the piston when it is changed into mechanical energy and this is in turn imparted to the flywheel when it is capable of doing continuous work.

If you have ever sawed a stick of cordwood

or done any other kind of labor with your hands, you know what *work* is and you also know that to do work takes *time* and *effort*. So if you wanted to measure the work you had done in hoeing turnips you would have to take into account the time you had spent and the effort you had made.

Now *mechanical* energy, or work, or power is measured in much the same way and that is by the number of pounds that a machine can raise a distance of one foot in one second.

For still greater convenience a *unit of work* was introduced by James Watt; it is known as *horse power*[1] and is written H. P. According to Watt a machine which developed one horse power could raise 550 pounds 1 foot in 1 second or as it is more commonly written

1 H.P. = 550 foot-pounds per second.

Since there are 60 seconds in a minute and since nearly all measurements of this kind are in minutes another way of writing it is

---

[1] A horse power, that is work done which will lift 33,000 pounds 1 foot high in one minute was obtained by Watt and Boulton, who took as their standard the strong dray horses that worked eight hours a day at the London breweries.

1 H.P. = 550 × 60 foot-pounds per minute, or
1 H.P. = 33,000 foot-pounds per minute.

And this is what the number means which you will use in the formula for calculating the horse power of a single cylinder engine as given below.

How to Calculate the H.P. of Your Engine.—To calculate the *approximate* horse power of a single cylinder engine use this formula:

$$\text{H.P.} = \frac{P \times L \times a \times 2R}{33,000}$$

where H.P. stands for the approximate horse power and is what you want to find,

P stands for the pressure of the steam on the piston[1] and is measured in pounds per square inch,

L stands for the length of the stroke in feet,

a stands for the *area* of the piston head in square inches and is found by multiplying the diameter of the piston by 3.1416,

2 R stands for twice the observed number of revolutions of the flywheel[2]—this meas-

---

[1] This is shown roughly by the pressure gauge of the boiler.
[2] This can be found by a *speed indicator*, a description of which is given in Appendix E.

Steam, the Giant Power

urement is the same as the piston speed, and

33,000 stands for the number of foot-pounds which equal one horse power.

Now in your horizontal engine which has a 1-inch cylinder and a 1¾-inch stroke (or 1.75), suppose that P (see above) is 40 pounds to the square inch and that you have found R to be 300 revolutions per minute.

Then: $L = \dfrac{1.75}{12} = .145$ feet

$a = \pi r^2 = 3.1416 \times .5^2 = 3.1416 \times .25 = .7854$ square inch.

$2R = 2 \times 300 = 600$ revolutions per minute.

Now substituting these values for the letters of the formula you have

$$H.P. = \dfrac{40 \times 145 \times 7854 \times 600}{33,000}$$

or $H.P. = \dfrac{2.733}{33,000}$

or $H.P. = .08 = \frac{1}{12}$ horse power.

How to Calculate the Size of a Boiler.—When I say the *size of a boiler,* I mean the *heating surface* of a boiler necessary for a stationary engine of known horse power.

The American Society of Mechanical Engineers have found by experiment that if 34.5 pounds of water are evaporated from and at 212 degrees Fahrenheit—that is, changed into steam, that 1 *boiler-horse power* is developed.

Now a boiler horse power is the amount of power which is necessary to run an engine rated at 1 *indicated*[1] horse power. It has also been determined that in order to evaporate 34.5 pounds of water from and at 212° F. the boiler must have 10 square feet of heating surface. That is to say, for every boiler horse power and likewise for every indicated horse power of the engine you must have 10 square feet of heating surface.

Now the heating surface of a boiler is the part that is actually heated by the fire. This is not only the part that the fire strikes directly but the whole surface of the fire-box sheet plus two-

---

[1] This can be found by using a *speed indicator.* See Appendix E.

thirds the area of the smoke-box sheet plus the combined area of all of the tubes.

In the language of a formula the heating surface needed in a boiler so that it will generate steam to run an engine of known horse power is

$$H.S. = I.H.P. \times 10$$

where H.S. = the heating surface and is what you want to find.

I.H.P. = the indicated or known horse power of the engine and

10 = the number of square feet of heating surface required to develop 1 boiler horse power.

Since, now, your engine develops $\frac{1}{12}$ or .08 of a horse power the

$$H.S. = .08 \times 10 \text{ or } .8 \text{ of a square foot}$$

of heating surface to run the engine to its fullest power.

To Calculate the Heating Surface of a Locomotive Boiler.—The amount of heating surface needed for a model locomotive boiler to develop 1 horse power, that is to evaporate 34.5 pounds of water from and at 212° F. is only 3 square feet.

Hence to calculate the number of square feet of heating surface needed for a locomotive boiler with engines of known horse power use the following formula:

$$H.S. = (H.P._1 + H.P._2) \times 3$$

where H.S. = the heating surface and is what you want to find,

H.P.$_2$ = the power of the other cylinder and

3 = the number of square feet of heating surface needed by the boiler to develop one boiler horse power.

CHAPTER X

A HOT AIR, OR CALORIC, ENGINE

The Parts of the Engine: Making the Engine; The Expansion Cylinder, The Transfer Piston, The Piston Rod, The Connecting Rod, The Power Cylinder, The Power Piston, The Connecting Rod, The Connecting Pipe, The Standards, The Crank and Crankshaft, The Fire-box—How to Operate the Engine—How the Engine Works.

The hot air engine is the easiest of all engines to build; it is the simplest to make, if we except the steam turbine, and it is by all odds the safest to use.

It is called a *hot air* engine because it is worked by the *expansion* and *contraction* of hot air—that is the air in the cylinders is first heated and then cooled and the same air is used over and over again. And it is truly surprising how very fast the temperature of the air can change.

It is also called a *caloric engine* in virtue of the fact that the word *caloric* means heat; this comes from the root *calor* which is Latin for heat and so naturally the term caloric engine is a more scientific name for that which the common people call a hot air engine.

The Parts of the Engine.—There are six principal parts to a hot air engine; namely, (1) the *displacement*, or *expansion cylinder* in which the air is heated and cooled; (2) the *displacement* or *transfer piston* with its *piston rod* and *connecting rod*; (3) the *power cylinder* where the working force is developed; (4) the *power connecting rod*; (5) the *crankshaft* with its *cranks, pulley* and *flywheel,* and (6) the base which holds the lamp or forms the furnace for developing the heat.

Making the Engine.—*The Expansion Cylinder.*—The expansion cylinder, as shown in Fig. 67, is the large one; it is made of a piece of iron pipe with a ⅛-inch thick wall, 2¼ inches in diameter outside measurement, and 4 inches long, and is threaded at both ends.

Get two caps threaded to fit the ends of the pipe and drill a 1-inch hole in the center of one of them and thread it; this cap is for the top head of the cylinder. A stuffing box can be fitted to the head and it will make it a little more efficient if you want to put one on.

A Hot Air, or Caloric, Engine

The Transfer Piston.—Before screwing the head caps on the cylinder, make the transfer pis-

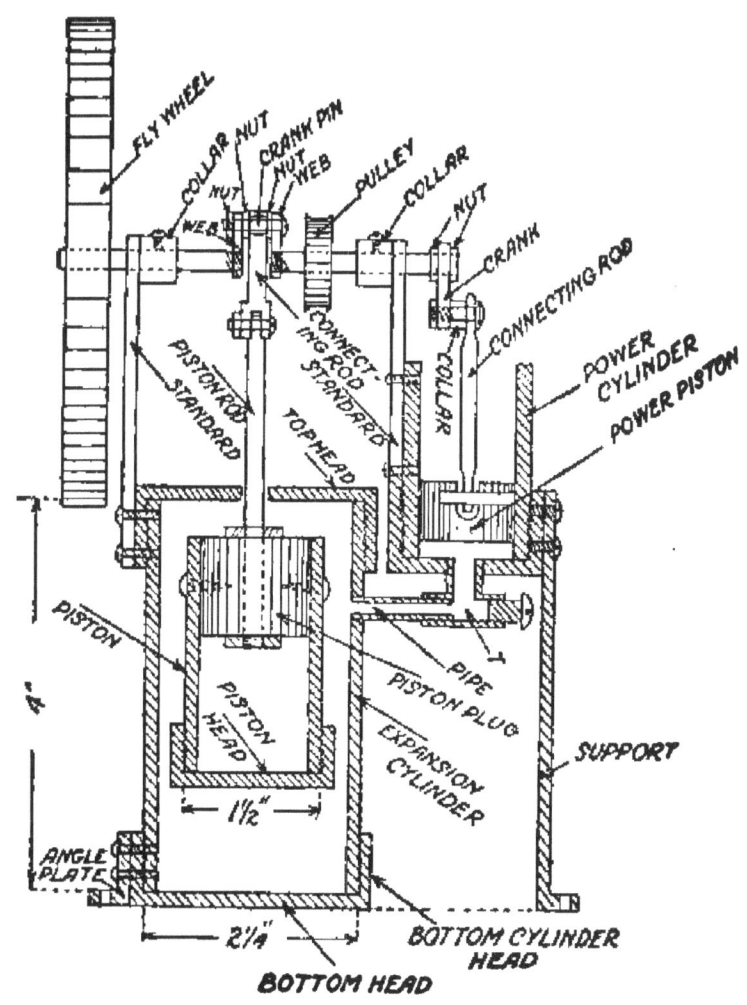

Fig. 67. Cross-Section of Hot-Air Engine

ton which moves in it. This can be a piece of pipe 1½ inches in diameter and 2⅜ inches long.

The reason it is made smaller than the cylinder will be explained presently. Thread one end of the pipe and screw a cap on it.

Drill two holes, $3/32$ inch in diameter, $1/2$ inch from the open end and opposite each other as shown in the cross-section view; then turn, or whittle out, a hard-wood plug that will fit the inside of the piston pipe tightly and make it 1 inch long and drill a $3/16$-inch hole through the middle of it.

The Piston Rod.—Make the piston rod for the transfer piston of brass or soft steel $3/16$ inch in diameter and $3\tfrac{3}{8}$ inch long, thread one end of it to a length of $1\tfrac{1}{4}$ inches, file the other end flat and drill a $1/8$-inch hole through it.

Screw a nut on the end of the piston down as far as it will go, put the threaded end through the hole in the piston and screw on another nut. You can now force the plug into the end of the pipe and then put in a couple of wood screws.

Slip the piston rod through the hole in the cylinder cap and screw the caps on the cylinder pipe with the piston inside of it.

The Connecting Rod.—This is the next part

to make. It is also made of a piece of $3/16$-inch rod; have it 1 inch long with a hole drilled in one end for a bearing for the crank pin; drill a hole into the other end and thread it so that the connecting fork can be screwed into it.

Make the connecting fork of a piece of $5/16$-inch rod; drill a hole through the fork for the bolt which is to form the pin; turn, or file, down the other end, thread it and screw it into the end of the connecting rod. The distance between the centers of the holes in the connecting rod is 1 inch.

The Power Cylinder.—This cylinder is half as large as the expansion cylinder and its piston must fit snugly and yet move easily in it.

Use a piece of iron pipe or brass tubing—the latter is the best—and thread one end of it. Drill a $1/4$-inch hole through a cap, which is threaded to fit the cylinder pipe, and thread it to take a $5/16$-inch pipe.

Screw the cap on the cylinder pipe and then drill two $3/32$-inch holes through one side of the cylinder, 1 inch apart, in a line with each other and thread them so that one of the standards which supports the crankshaft can be screwed to it. Of course the screws must not project through the inside of the cylinder.

The Power Piston.—Whatever kind of metal you use for the power cylinder make the piston of the same metal so that the expansion of each will be about the same.

The inside of the cylinder must be quite true and smooth and the piston must fit it pretty accurately, as it is not packed. Make the piston ⅝ inches long, cut a groove by drilling a hole ⅛ inch in diameter, ¼ inch from one of the ends; saw two cuts ⅛ inch apart through the end until it meets the hole and you will have a groove wide enough to set in the end of the connecting rod.

Drill a ³⁄₃₂-inch hole through the piston at right angles to the slot so that a pin can be driven in to pivot it to the connecting rod.

The Power Connecting Rod.—There is no piston rod used to link the piston of the power cylinder to the crank of the crankshaft, but instead a connecting rod only is used.

For the connecting rod get a rod ³⁄₁₆ inch in diameter and 2¼ inches long; hammer both ends flat, file them up smooth and drill a hole in each end with their centers 1⅞ inches apart. Set one end of the connecting rod in the groove in the piston and pivot them by driving in a pin.

The Connecting Pipe.—All is now ready to connect the power cylinder to the expansion cylinder. The easiest way is to cut off a piece of $5/16$-inch brass pipe 1 inch long and thread both ends of it.

The ends of a T,[1] as you know, are threaded on the inside, but in this case you want the outside of the lower end of the T threaded. Screw the stem of the T into the hole in the head of the cylinder, screw a screw into one end of the T, screw one end of the pipe into the other end of the T and screw the other end of the pipe into the hole near the top of the expansion cylinder.

The Standards.—The standards which support and form the bearings for the crank, crank shaft and flywheel are brass bars $1/8$ inch thick and $3/8$ inch wide and one of them is $2\frac{1}{2}$ inches long and the other one is $3\frac{3}{8}$ inches long.

Drill a $3/32$-inch hole in one end of both standards for the crankshaft; in the other end of the short standard drill two $1/8$-inch holes, 1 inch apart, and in the other end of the long standard

[1] See Chapter VI.

drill two holes ⅝ inch apart. Screw the short standard to the power cylinder but don't screw the long standard to the top of the expansion cylinder just yet.

The Crankshaft and Crank.—The crankshaft can be formed of a single length of bent rod or it can be made of two pieces of rod each of which is 3/16 inch in diameter and 1⅞ inches long, and thread it at both ends.

Make two *webs,* as the arms for the middle crank are called, of ⅛ inch thick brass, or iron, strip and have each one ⅜ inch wide and ⅝ inch long. Drill a 3/16-inch hole in one end of each web and a ⅛-inch hole in the other end of each one; thread all of them and screw a web to each of the shafts.

Fix a pulley on one of the shafts and slip a collar fitted with a set screw over the end of each shaft. Now push the end of this half of the shaft through on a nut.

Next, make a crank of a strip of ⅛ inch thick brass, or iron, ⅜ inch wide and ⅞ inch long; drill a 3/16-inch hole in one end and a ⅛-inch hole in the other end; put it over the end of the shaft close up to the nut, screw on another nut and tighten it up so that the crank can't possibly slip on the shaft.

Couple the connecting rod to the crank and keep them apart by means of a thick washer, or a collar, so that there will be enough clearance for the rod to pass across the end of the crankshaft without danger of striking it.

Slip a collar with a set screw in it over the end of the other shaft, put the latter through the hole in the long standard and screw the standard to the side of the expansion cylinder.

Put a ⅛-inch screw through one of the webs of the middle crank for the crank pin, screw on a nut, slip a *sleeve,* that is a thin piece of metal tubing, over the screw to form a smooth bearing and put the end of the connecting rod over it.

Screw on another nut, then screw on the other web and finally screw a nut on the end of the screw good and tight so that the two shafts are in *alignment,* that is in a straight line, and so that the webs can't slip and a rigid crankshaft is formed.

Now slip another collar with a screw in it over the end of the crankshaft, put the end of the long standard over it, screw the standard to the side of the expansion cylinder; screw the collar up close to it and then *key,* or otherwise fix to the end of the crankshaft a heavy 5- or 6-inch flywheel.

The Fire-Box.—There are four parts to the fire-box and these are the top, the body and the

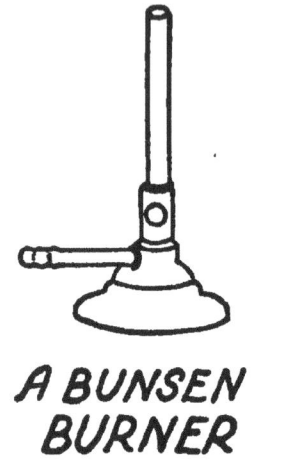

A BUNSEN BURNER
Fig. 68. A Bunsen Burner

base and the support. The top and base can be cut out of heavy sheet metal and the body shaped up of sheet iron, but it is easier to make a cast iron fire-box and it is better because it is more rigid and as the engine sets on top of it a firm support is necessary.

The top should be ⅛ inch thick, 5⅛ inches in diameter and have a hole in it 2½ inches in diame-

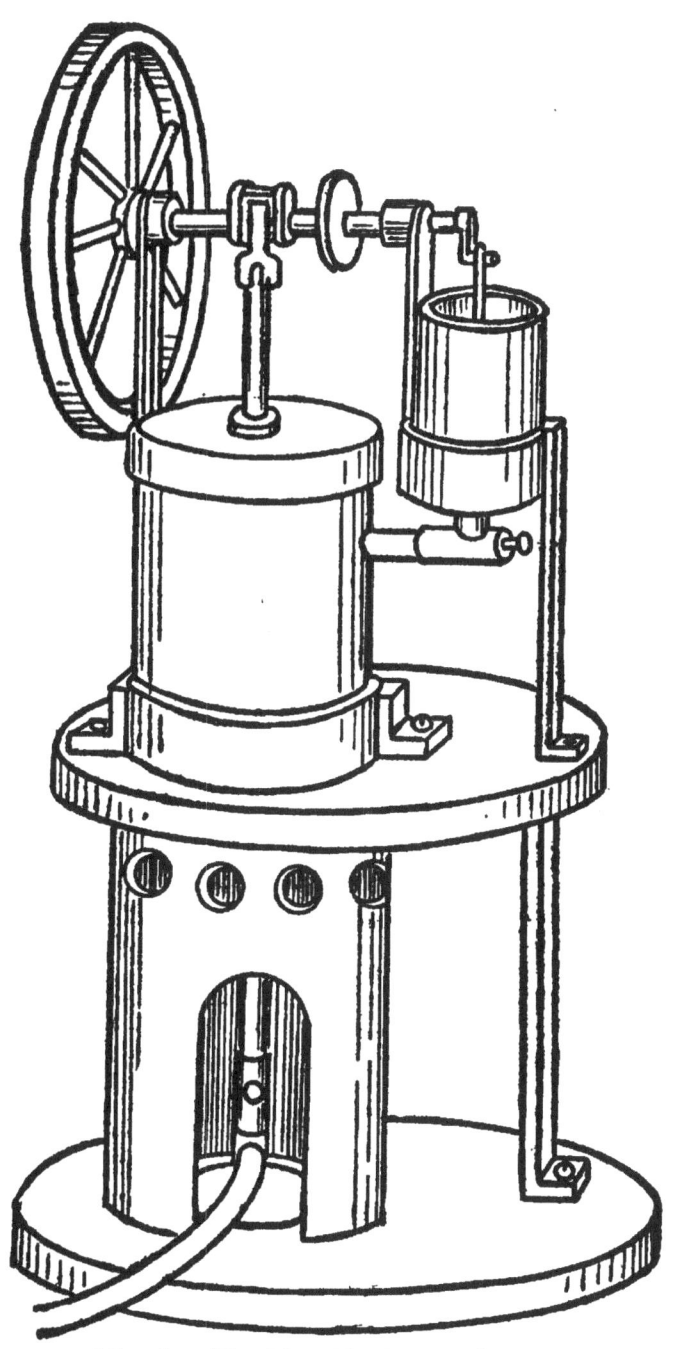

Fig. 69. The Hot Air Engine Complete

ter with its center ⅞ inch from the true center of the top. The foot and the support of the engine is screwed to the top of the fire-box with the lower end of the expansion cylinder setting in the hole.

The body can be about 4 inches in diameter and an inch or so higher than the Bunsen burner, see Fig. 68, you use to heat it. It must have a row of holes in the top and bottom to give the air a chance to circulate freely. The bottom is simply a disk of cast iron, being ¼ inch thick and 5½ inches in diameter. The hot air engine is shown assembled in Fig. 69.

To Operate the Engine.—Set a Bunsen burner in the base, or an alcohol lamp will work it in a pinch, light it and see to it that the flame is in the center of the bottom of the expansion cylinder. Give the flywheel a couple of turns to start it going and it will begin to develop power.

How the Hot Air Engine Works.—To the end that you may know without having to rack your brain too hard just how a hot air engine works I have drawn a simple diagram of it and this is shown in Fig. 70.

When the bottom of the expansion cylinder is

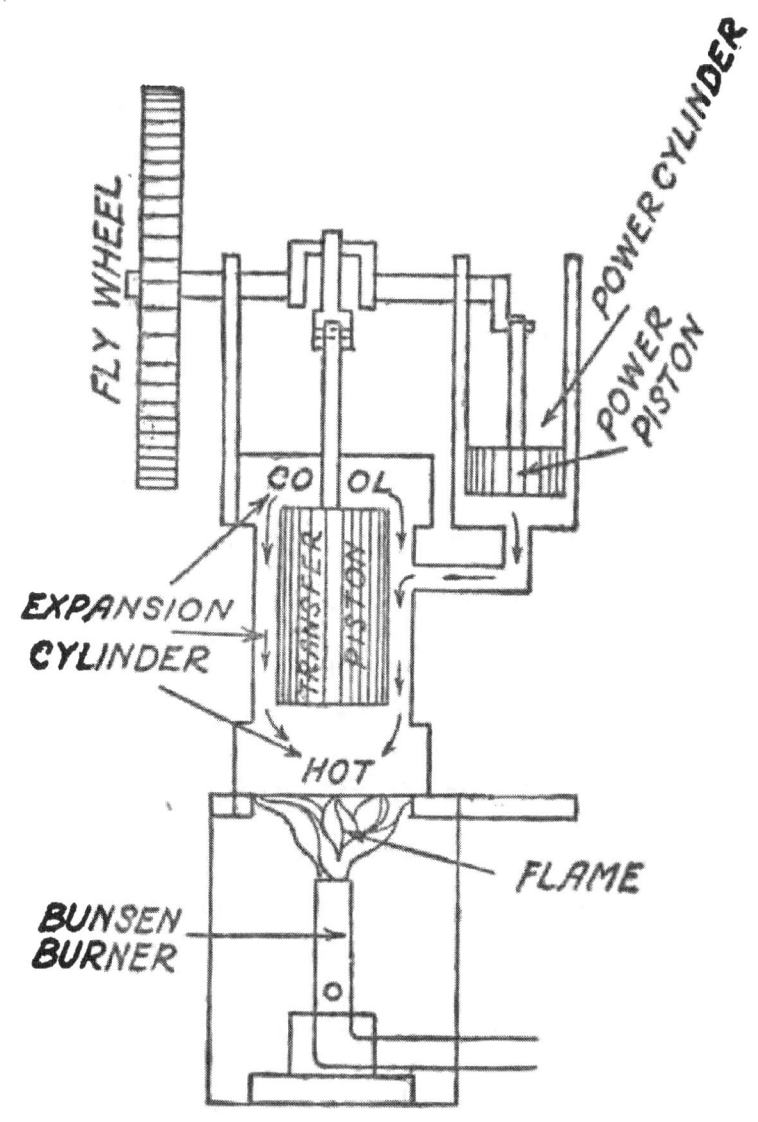

Fig. 70. How the Hot Air Engine Works

heated and you start the engine off by turning over the flywheel the loose fitting transfer piston travels down and displaces, or transfers, the hot air in the bottom of the cylinder into the top part of the latter. This is easily done as there is plenty of room between the piston and the cylinder for the air to move in.

Not only is the hot air forced up into the top of the expansion cylinder, but it is also drawn into the power cylinder, for, while the transfer piston is down, the power piston is going up. But the moment the expanding hot air reaches the upper part of the cylinder and fills the power cylinder it cools off; this of course makes it contract and having lost its heat energy the power piston is forced down by the pressure of the air outside on it, which is 15 pounds to the square inch.

As the power piston is being forced down by atmospheric pressure the expansion piston is being raised and this pushes the cooled off air in the top part of the expansion cylinder into the lower part of it when it is heated again.

Just remember that the only purpose of the transfer piston is to move the air up and down

A Hot Air, or Caloric, Engine 213

in the cylinder from the hot to the cold end and back again. Also bear in mind that the same air is used over and over again. The purpose of the screw in the end of the T is to clean out the pipe connecting the cylinders should it get clogged up.

A hot air engine is a very efficient machine for changing the energy of heat into mechanical motion and it is a good one for you to build and use. The reason large engines of this kind are not used is because they must be very large when compared to other types of engines developing the same power and, besides, the heat very quickly burns out the bottoms of the cylinders.

You can make a hot air engine much smaller than the one I have given or very considerably larger if you want it to do real work, providing you hold to about the above proportions.

CHAPTER XI

A ⅛-H. P. GAS ENGINE

Gas Engines Versus Steam Engines—The Parts of a Gas Engine: The Cylinder—The Inlet and Exhaust Valves; The Exhaust Valve Mechanism; The Igniter. The Camshaft Bearings; The Cam and Camshaft; The Timing Gears; The Piston; The Connecting Rod; Making the Crankshaft; Assembling the Crankshaft; The Bed of the Engine; Assembling the Engine; About Oiling the Engine; The Flywheel and Pulley—How the Gas Engine Works.

And now we come to the last of our model machines for converting heat into mechanical motion and this is the *gas engine*.

While the gas engine stands second to the steam engine as a giant power producer, it ranks ahead of it in modern achievement for it not only made the automobile practicable but it made the airplane possible.

Now a gas engine differs from a steam engine in that the power of the first is produced directly inside of the cylinder by the explosion of a gas while with a steam engine the power must be developed, as you well know, in a separate and distinct apparatus which we call the boiler.

Again a gas engine is different from a steam engine in that in the former there is only one power stroke to every four movements of the piston, the other three strokes depending on the *momentum* of a heavy flywheel where only one cylinder is used, while in the latter every to and fro movement of the piston is a power stroke.

The third great difference between a gas engine and a steam engine is that, as you have seen, the explosion of gas which develops the power stroke is nearly instantaneous, whereas the latent heat in steam causes it to expand gradually and to develop power continuously.

For these reasons a gas engine only reaches its full working power when it is running at its highest speed and, hence, any sudden attempt to use it will *stall* it, whereas the power of a steam engine can be used from the moment the throttle is opened and steam enters the cylinder.

One good feature about a gas engine, though, is that it can be built easier and cheaper than a steam engine and boiler of equal power; another point in its favor is that it takes up less room and weighs less for the power developed, and, third, it requires far less attention after it is started.

The Parts of a Gas Engine.—There are five main parts to this gas engine and these are (1) the *cylinder;* (2) the *piston* with its *connecting rod;* (3) the *crankshaft;* (4) the *flywheel* and (5) the *base.*

Then there are five auxiliary, or smaller, parts, but these are just as important as the main parts, in fact they are the very vitals of the engine. Named these parts are: (1) the *inlet valve;* (2) the *exhaust valve;* (3) the *camshaft* and *cam;* (4) the *timing gears,* and (5) the *igniter.*

The Cylinder.—This can be made of an iron pipe but it must be bored out true and smooth. Give it an inside diameter of 1 inch, make it 2¾ inches long and thread one end of it as shown in Fig 71.

Get a screw cap to fit it and drill a ¼-inch hole through the opposite sides of the wall, drill another ¼-inch hole through the center of the head of the cap and thread all of them to fit a 5⁄16-inch pipe. The first two holes are for the inlet and the exhaust valves which let in the fresh gas and

Fig. 71. Side View of the Gas Engine

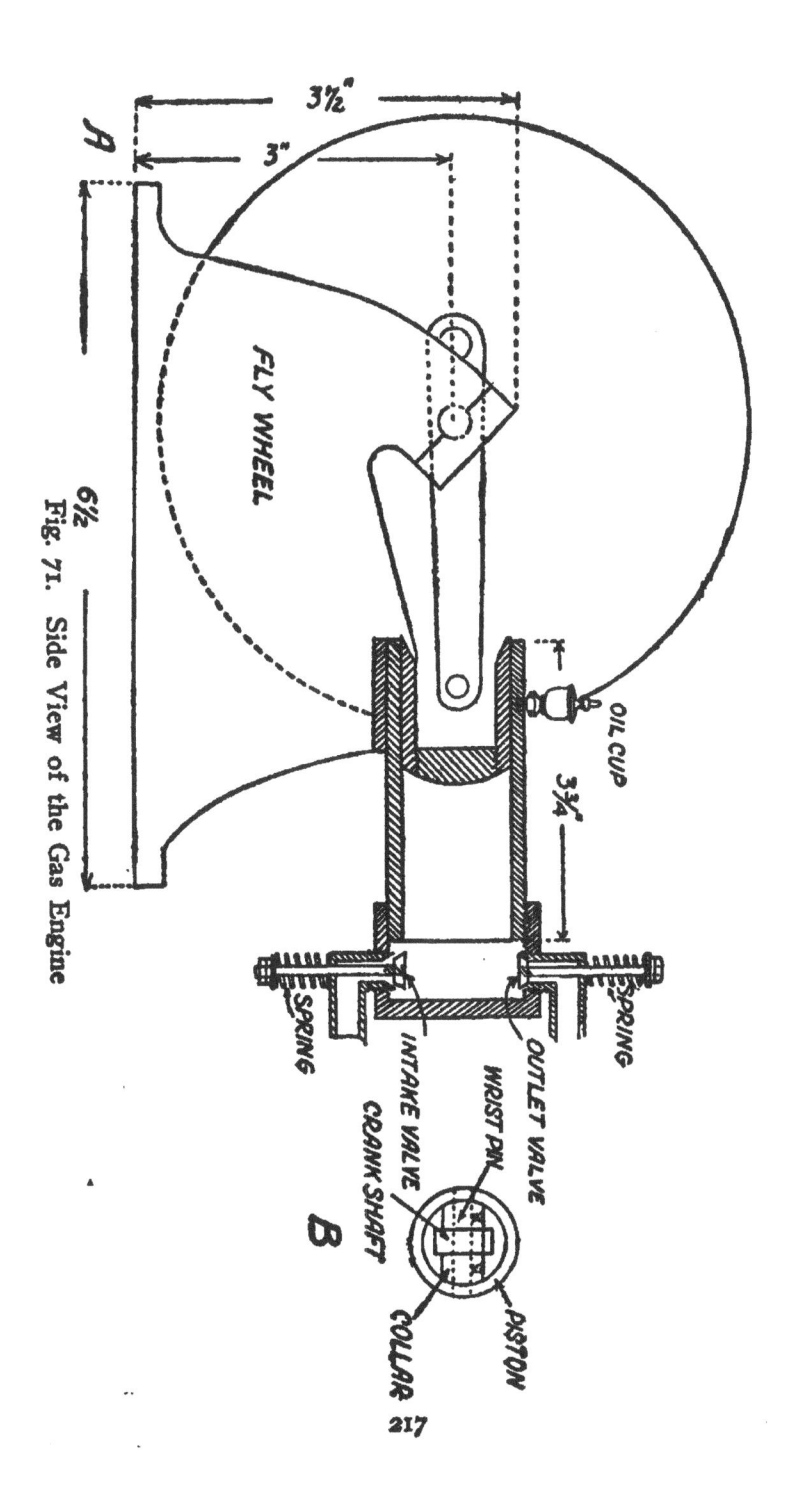

let out the burned gases. The hole in the head of the cap is for the igniter.

Drill a $\frac{1}{8}$-inch hole $\frac{5}{8}$ inch back from the front end of the cylinder and through the wall and thread it for an oil cup. These can be bought ready made with screw tops for 25 cents each.

The Inlet and Exhaust Valves.—A simple way to make both the inlet and the exhaust valves, if you haven't a lathe, is to take a $\frac{5}{16}$-inch elbow for each one (see Chapter VI).

Thread one of the ends on the outside and then ream it out to form a beveled edge as shown in Figs. 71 and 72, to form a seat, as it is called. Drill a $\frac{1}{16}$-inch hole through the elbow for the *valve stem* so that it will be exactly in the center of the end of the elbow that is beveled out.

Make a *valve head* of a disk of soft steel $\frac{1}{16}$ inch thick and $\frac{3}{8}$ inch in diameter and bevel the rim so that it will fit accurately in the beveled end of the elbow. Drill a $\frac{1}{16}$-inch hole through the center of it and thread it. For the valve stem use a soft steel rod $\frac{1}{16}$ inch in diameter and $1\frac{3}{8}$ inch long and thread both ends of it, screw one end into the valve head as shown at A in Fig.

A ⅛-H. P. Gas Engine 219

71 and slip the other end through the hole in the elbow.

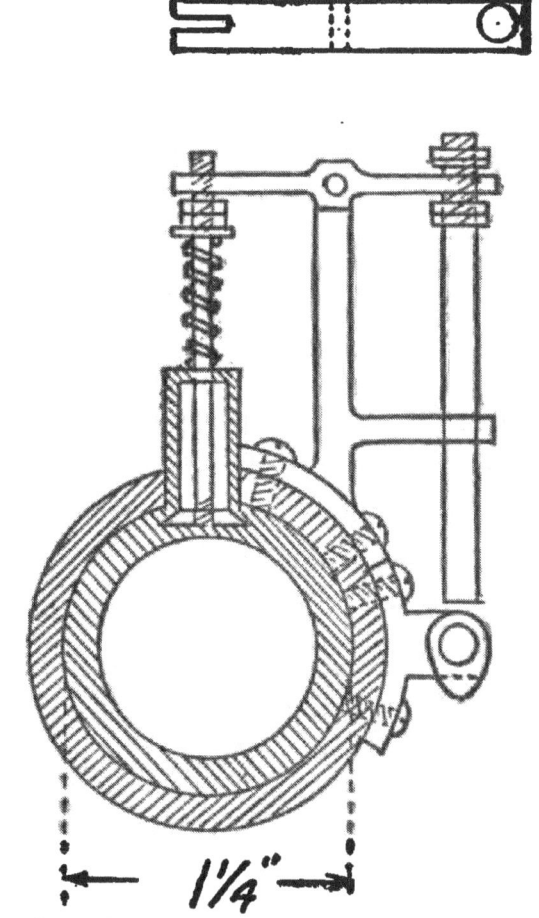

Fig. 72. Cross-Section View of Cylinder Showing Exhaust Mechanism

Next make an open *spiral spring* of very thin brass wire ¹⁄₁₆ inch in diameter. Put this on over the valve stem with one of its ends resting against the elbow, slip on a washer and then screw two nuts on the end of the stem. The purpose of the spring is to make the valve head seat properly.

The inlet valve is the lower one and no mechanism is needed to work it, for when the *suction stroke* takes place, that is, the stroke of the piston which pulls the explosive mixture of air and gas into the cylinder the suction is sufficient in itself to lift the valve out of its seat.

The Exhaust Valve Mechanism.—The elbow, valve and valve stem of the exhaust valve are made exactly like the inlet valve, but it must be mechanically opened against the pressure of the spring to let out the burnt gases.

A picture of this device to open the valve is shown in the cross section view Fig. 72 and in the top view of the engine in Fig. 72, while all the parts drawn out in perspective are shown at A, B, C and D in Fig. 73. It is easier to make a pattern for the *support* B and the *rocker arm* D and have them cast in brass than to shape them up by hand.

The support is formed of a *standard* 1¼ inches high at the back and 1⅜ inches high in front; half of the top end is cut away as shown at B

A ⅛-H. P. Gas Engine 221

and a hole is drilled through the remaining part. The standard rests on a *base* curved to fit the cylinder and you must drill a hole in each end for the screws. Projecting out in front of the

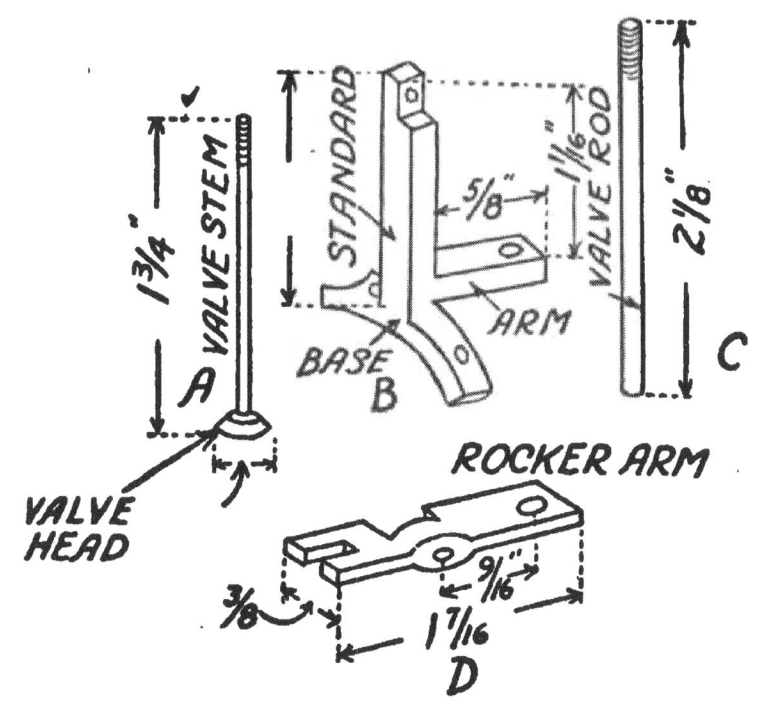

Fig. 73. Parts of the Exhaust Valve Mechanism

standard is an arm with a hole in the end just large enough to let the valve rod, see C in Fig. 73, slip through it.

The rocker arm D is 1/16 inch thick, except near the middle where it bulges out to 3/16 inch;

it is ⅝ inch wide and 1⁷⁄₁₆ inch long and a slot ¹⁄₁₆ inch wide and ¼ inch long is sawed out of one end.

Half of the bulged part is cut away and a hole is drilled through the part that is left. A ⅛-inch hole is drilled in the other end and the distance between the centers of these holes is ⁹⁄₁₆ of an

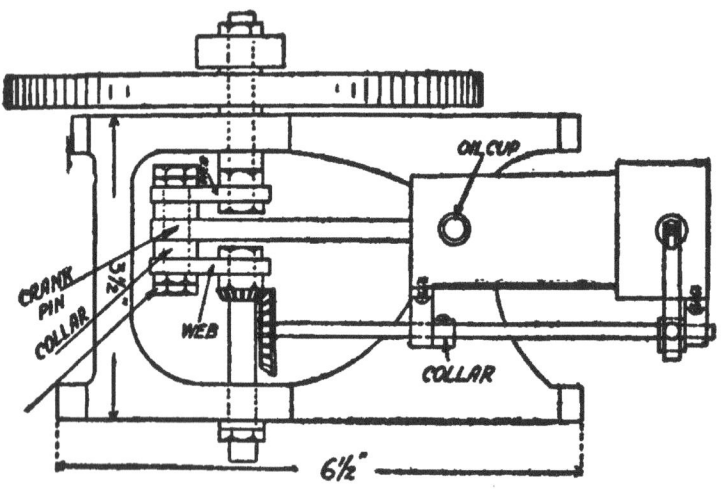

Fig. 74. Top View of the Engine

inch. The rocker arm is now pivoted to the standard and, since both are cut out half way, the hole in the end of the rocker arm and the hole in the end of the arm of the support will be directly over and in a line with each other.

The cam rod is a steel rod ⅛ inch in diameter and 2⅛ inches long and threaded on one end;

screw on two nuts, slip the end of the rod through the hole in the arm of the standard and the end with the nuts on it through the hole in the rocker arm; then screw on two more nuts as shown in the cross section view, Fig. 72.

When you have assembled the parts, screw the base of the support to the cylinder so that one end of the base, which should be rounded out, sets against the elbow as is also shown in Fig. 72, when the slotted end of the rocker arm will set over the end of the valve stem.

The Igniter.—There are two ways to fire the fuel mixture of a small internal combustion engine: (1) by a *hot metal tube* and (2) by an *electric spark*.[1] As the hot tube igniter is the simplest to make, as well as to use, I prefer it for this little engine.

The whole igniter must be made of iron, except the hot tube which is of steel and the *mica* washers and rings which insulate the tube when hot from the shell of the igniter to keep it from losing too much of its heat.

[1] For a very full description of how electric spark ignition works, see "Keeping up with Your Motor Car," by the author, and published by D. Appleton and Co., New York.

Get a piece of pipe 1¼ inches outside diameter and 1⅛ inches long and thread both ends of it; drill a 3/16-inch hole through the wall half way between the ends and thread it to fit a ¼-inch pipe. Fit two caps to the ends of the large pipe and drill a ½-inch hole in the center of one of them and also drill four screw holes around the hole in the center so that the cap can be screwed to the head of the cylinder, as shown in Fig. 75. Drill a 3/32-inch hole in the center of the other cap, thread it and put in a screw with a nut on it. Next get a large iron washer, 3/16 inch thick, that will fit inside the iron tube snugly and it must have a ½-inch hole in it; also get a small iron disk 1/16 inch thick and ½ an inch in diameter.

For the ignition tube use a piece of steel tube 5/16 inch in diameter and 1 inch long and, last of all, get a *mica* [1] washer, or use enough of them to make a washer 1/16 inch thick; two mica rings to fit over the steel tube—these are shown by the black parts in Fig. 75—and a mica disk each of which is 1/16 inch thick and ½ inch in diameter.

[1] Mica can be bought in hardware stores. Mica in every shape and form is sold by Eugene Munsell and Co., 68 Church Street, New York City.

A ⅛-H. P. Gas Engine 225

To assemble the igniter screw the cap with the large hole in it to the head of the cylinder; next screw in the iron pipe; set the mica washer

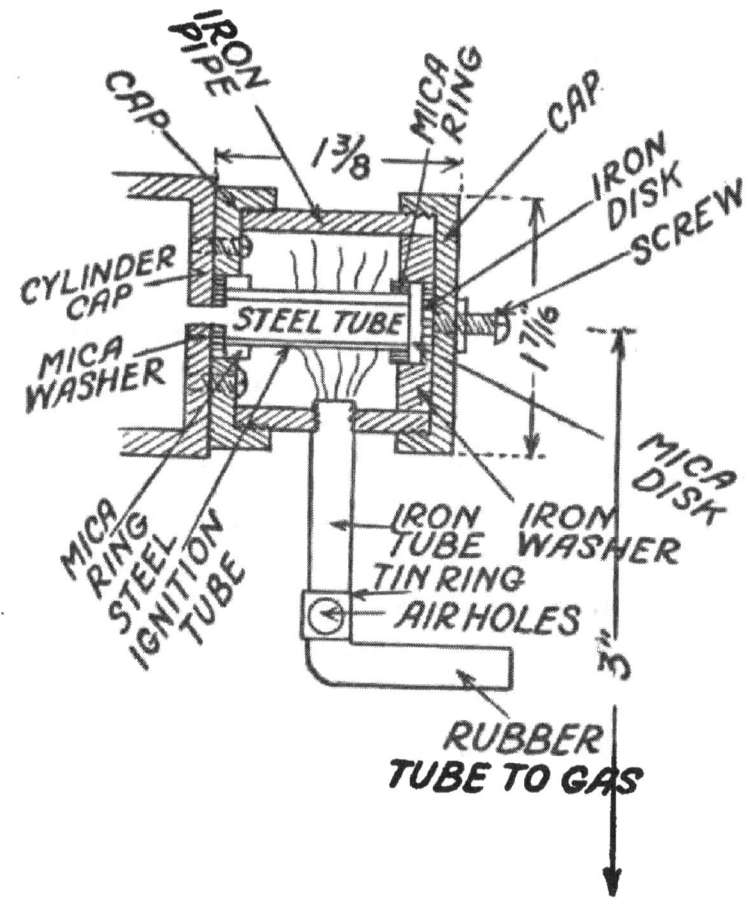

Fig. 75. Cross Section View of the Igniter

against the head of the cylinder; slip the mica ring over one end of the steel tube and set the tube up against the washer.

Slip a mica ring on the other end of the steel tube and fit the iron washer over the mica ring and inside the pipe; set the mica disk against the end of the tube and inside of the iron washer; also set the thin iron disk against the mica disk inside of the washer and then screw on the remaining cap.

Finally, tighten up the screw in the cap to make the steel tube gas-tight around the ends. This completes the igniter except for the Bunsen burner which is used to heat the steel ignition tube. It is called a *Bunsen burner* because it was invented by Bunsen, a great German scientist who lived in the 19th century.

It is formed of a tube the lower end of which is connected to a supply of gas and the upper end is left open; holes are drilled in the pipe near the lower end and, when the gas is lit at the upper end, air is drawn into the pipe where it mixes with the gas and this makes a hot flame.

All you have to do to make a Bunsen burner for the igniter is to get a ¼-inch iron pipe 2½ inches long, thread one end of it and bend the other end over about 1 inch; drill a ⅛-inch hole clear through the pipe just above the bend and fit a tin, or a brass, ring ⅜ inch high and which also has a pair of holes drilled in it, over the pipe where the air holes are. By turning the ring around, the air holes are opened or closed and hence the amount of air can be regulated.

Screw the Bunsen burner pipe into the iron pipe of the igniter and all of your fine work on the engine is done.

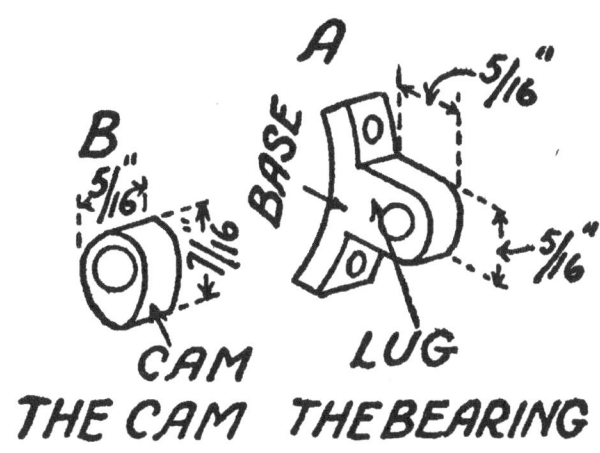

Fig. 76. The Cam and Camshaft Bearing

The Camshaft Bearings.—Make two patterns as shown at A in Fig. 76 for the bearings for the camshaft and have the base of one ⅛ inch thick and the base of the other ¼ inch thick; the reason one must be made thicker than the other is because one is screwed to the front end of the cylinder which is smaller all around by ⅛ inch than the cap to which the other bearing is screwed. The base of each bearing is ¾ inch long and each must be curved to fit the cylinder.

Glue a lug on the rounded side of each base ¼ inch thick, 5/16 inch wide and 5/16 inch long and drill a 3/16-inch hole through the end. After the patterns are cast, screw the one with the thinnest base to the extreme end of the cap on the cylinder and screw the other bearing to the front end of the cylinder as shown in the top view Fig. 74 and in the perspective drawing Fig. 79, be sure to have the holes in a straight line with each other.

The Cam and Camshaft.—To raise the cam rod up so that the rocker arm will push the valve stem down and thus open the exhaust valve when the exhaust stroke takes place, a *cam* is used and this is driven from the crankshaft by means of a camshaft and a pair of timing gears as shown in the top view, Fig. 74, and in the perspective drawing, Fig. 79.

The cam is an elliptic-shaped piece of steel with a hole through the large end as shown at B in Fig. 76, and its purpose is to raise the cam rod every time it turns around once, which it does since the lower end of the rod rests on it. It is screwed, keyed or otherwise fixed, on one end of the *camshaft*.

The camshaft is a steel rod ⅜₆ inch in diameter and 5¼ inches long. Slip the free end through the bearing on the cap of the cylinder, put a collar on it and then push it on through the bearing on the front end of the cylinder and screw up the collar so that the shaft can't slide in its bearings. A *timing* gear is fixed to the free end of the shaft.

The Timing Gears.—Now since there is only one explosion to every four strokes of the piston it must be clear that there is only need for the exhaust valve to open once in this number of strokes to get rid of the burnt gases.

To do this the camshaft must make only one revolution while the crankshaft makes two complete turns. To get this result a pair of *beveled gears* is used and the gear on the camshaft must have *twice as many teeth* cut on it as the one on the crankshaft which drives it.

A pair of *beveled gears,* one of which is ½ inch in diameter and has 15 teeth and the other 1 inch in diameter and with 30 teeth, is well suited to this engine. Bevel gears of these sizes can be bought of dealers in model supplies for about 75 cents per pair. The large bevel gear is, of course, fixed to the end of the camshaft and the small gear is keyed to the crankshaft as shown in Fig. 74.

The Piston.—Having all these small but highly important details attended to, you can now go ahead with the heavier parts of the engine.

The piston should be made of iron and carefully fitted to the inside of the cylinder; it is 1 inch in diameter and 1⅜ inches long. Make it of a piece of iron pipe 1 5/16 inches long and it ought to be turned down to make a good fit; thread the inside of one end and bevel the inside of the other end to allow room for the connecting rod. Make a screw plug 5/16 inch thick and thread it to fit the pipe and then round off the end as shown in Fig. 71.

Drill a 3/16-inch hole through the piston wall, ½ an inch from the open end for the *wrist pin,* or *gudgeon pin* as the pin that couples the connecting rod to the piston is more properly called. How the connecting rod is fitted to it will be told below.

The Connecting Rod.—Make the connecting rod of brass ¼ inch thick, 3/16 inch wide on one end and taper it down to ⅜ inch wide on the other end; let it be 3⅝ inches long and round off both ends as is also shown in Fig. 71.

In the small end drill a ⅜₁₆-inch hole for the wrist pin and in the large end drill a ¼-inch hole for the pin of the crankshaft to go through. You can make the connecting rod in two ways and these are (1) by having it cast in brass and (2) by shaping it out of a brass bar.

If you have it cast, make the pattern with bosses on the ends, that is projecting disks; but if you shape it out of a brass bar make two collars ⁷⁄₁₆ inch in diameter and ¼ inch long; drill a ⅜₁₆-inch hole through each one and round it off so that it fits the curve of the inside of the piston.

Get a piece of steel rod ⅜₁₆ inch in diameter, 1 inch long, for the wrist pin. Now push the end of the pin through the wall of the piston, on through a collar, next through the small end of the connecting rod, then on through the other collar and, finally, through the wall of the piston. There is no need to fasten the wrist pin in tight because, when the piston is in the cylinder, it is there to stay.

Making the Crankshaft.—This can be forged

of a single length of steel rod but it will be easier for you to make it of three pieces of steel and connect them together with brass, or steel, webs.

To make a crankshaft after the latter fashion get a piece of ¼-inch steel rod 1½ inches long for the crank pin and two more pieces of the same diameter and each of which are 2⅛ inches

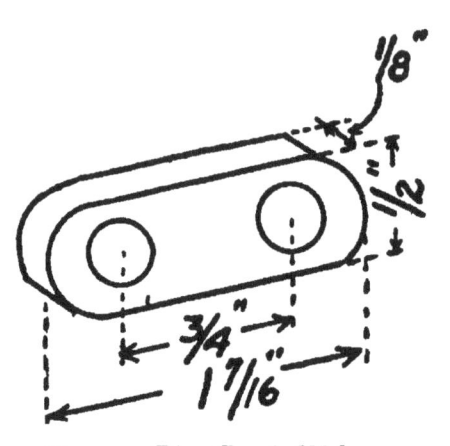

Fig. 77. The Crank Web

long for the shaft. Thread both ends of all the pieces and fit them with nuts.

Cut off two pieces of brass bar each of which is ⅛ inch thick, ½ an inch wide and 1⁷⁄₁₆ inches long for the webs and drill a ⁹⁄₁₆-inch hole in one end of each web and thread it to fit the shaft. Drill a ¼-inch hole in the other end of the web and have the centers of the holes exactly ¾ inch apart as shown in Fig. 77.

Assembling the Crankshaft.—The crankshaft can be completely assembled and the connecting rod coupled to it before it is mounted in the journal-bearing as the latter is *halved,* that is cut in two as shown in Fig. 71.

To assemble the crankshaft you will need ten nuts, three washers, 3/32 inch thick, and three collars, 3/16 inch thick; the washers and the collars must fit snugly over the shafts and all of them should be ½ inch in diameter.

Begin by screwing, or otherwise fixing, the small bevel gear to one of the shafts ⅝ inch down from the end, see Fig. 74, now screw on a nut, next screw on the web and, lastly, screw on another nut. On the other shaft screw a nut 7/16 inch down on one end, screw on the web, then screw on another nut and slip on a collar. The webs must be put on tight or they will twist around on the shafts.

Slip the end of the connecting rod over the crank pin, put a collar on each side of it, screw the ends of the webs down on the pin and screw two nuts on each end of it. This construction secures the crank pin to the shafts so that a fairly rigid unit is made of it, while the bearing of the connecting rod can turn freely.

The Bed of the Engine.—The bed for this en-

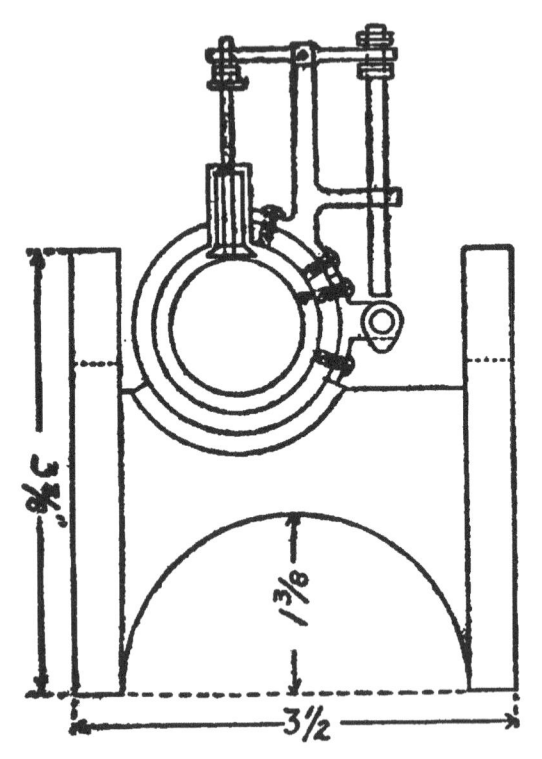

Fig. 78. Back End View of The Gas Engine

gine ought, by all means, to be a cast one. To make a pattern for the bed measure up the side, top and end elevations shown in Figs. 71, 74 and 78 and then by looking at the picture of the completed engine in Fig. 79 you will see precisely how it is put together.

If you have the bed cast in iron the journal boxes should be fitted with *Babbitt metal* bearings [1] for the friction of steel sliding on iron or brass is much greater than steel on Babbitt metal.

Assembling the Engine.—Drill four ⅛-inch holes in the *cradle* of the bed, that is the concave part on which the cylinder rests; drill four corresponding holes through the wall of the cylinder near the open end and screw the bed to the cylinder.

Put the piston in the cylinder and set the crankshaft in the bearings of the bed when the small timing gear on the crankshaft will mesh with the large timing gear on the camshaft as shown in Fig. 74 and this will bring the collar on the other end of the camshaft up against the journal box.

When you have the connecting rod, the timing gears and the crankshaft adjusted so that they all run smoothly, screw on the tops of the journal boxes. Slip a washer on each end of the crank-

[1] See Appendix.

shaft outside of the journal boxes and screw a nut on the end of the shaft that carries the bevel gear.

The Flywheel and Pulley.—The *flywheel* should be a heavy one with at least a ⅜-inch face and it should be 5 inches in diameter. A spoked wheel of this size can be bought ready to use.

It is a good scheme to fit a handle to one of the spokes to crank the engine with when starting it. Fix the flywheel on the end of the crankshaft outside of the journal box and slip on a washer.

The pulley can have a flat face ⅜ inch wide and a diameter of about 1 inch; fix it to the shaft next to the washer and screw on a nut all of which is shown in Fig. 74.

About Oiling the Engine.—One of the chief things to do to make an engine run and to keep it running is to supply all the working parts plentifully with oil.

The piston should be well lubricated with a good grade of *light* automobile engine oil and this can be done by remembering to keep the oil cup filled with oil all the time.

It is a good scheme to fit an oil cup in the end

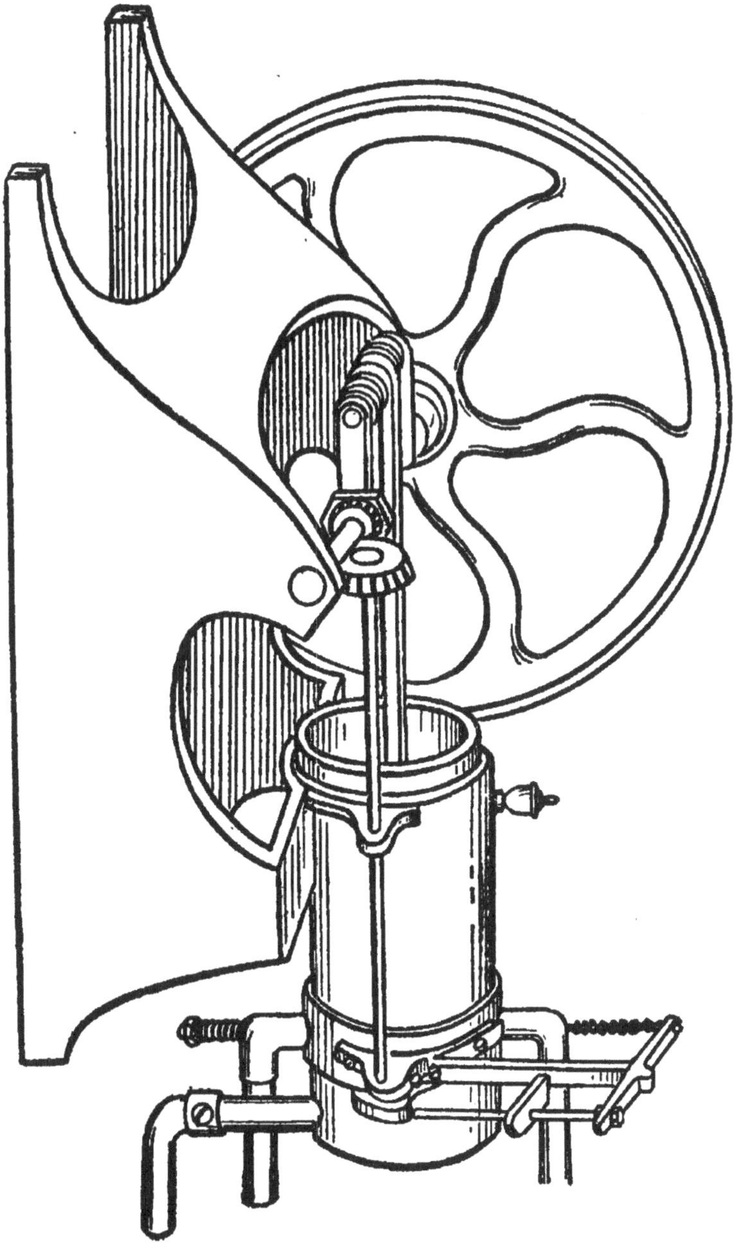

Fig. 79. The Model Gas Engine Complete

of the connecting rod and also in each bearing of the crankshaft and don't forget to oil up all the other little parts often.

How the Gas Engine Works.—The way a single cylinder gas engine works can be easily understood by looking at the diagrams shown in Fig. 80.

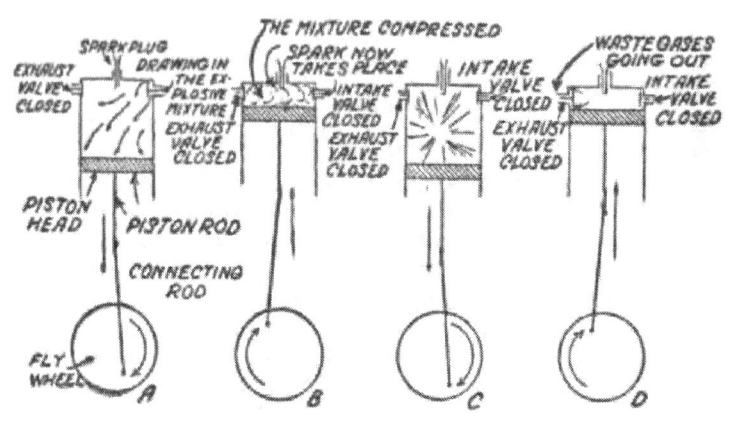

Fig. 80. How the Gas Engine Works

The diagrams A, B, C and D show the same cylinder during four strokes of the piston or *cycles* as they are called, and, hence, during two complete revolutions of the crankshaft.

The diagram A shows the piston making its *suction stroke* and this lifts the inlet valve out of its seat and draws in the mixed air and gas, or fuel *mixture*. While this operation is taking place the exhaust valve is kept closed by its spring.

When the piston has reached the end of its suction stroke and begins to move back it *compresses* the fuel mixture, or *charge* as it is now called, and this in consequence is named the *compression* stroke. Of course the inlet and the exhaust valves are closed while the compression stroke is being made, not only by their respective springs but because the pressure of the gas is on them.

The fuel charge is not fired by the hot tube of the igniter until the gas is compressed for the reason that it is not hot enough; but when a gas is compressed it develops heat and this, added to the heat of the hot tube, raises the temperature high enough to explode the gas.

The instant the gas is fired it explodes and the force of it drives the piston out and so makes the *power stroke*. Naturally the force of the explosion acting on the valves would hold them in their seats even if the springs did not keep them closed.

Now, while the piston was making these three strokes the camshaft made only three-quarters of a revolution but at the end of the power stroke the cam on the camshaft reached a position where it began to lift the exhaust valve rod and this in turn opened the exhaust valve; then as the piston moved back it forced the waste gases out of the cylinder through the exhaust port.

From what has been said above it will be seen that there are four separate operations performed by the piston and for this reason a gas engine of this kind is called a *four cycle* engine; it is also called an Otto cycle gas engine from Otto, a German engineer who invented it.

Since there is only one power stroke to every two revolutions of the crankshaft, a heavy flywheel must be used to carry the crankshaft around during the other three strokes. In automobiles, motor boats, motor cycles and airplanes four or more cylinders are used and the pistons of all of them are connected to one crankshaft in such a way that the power strokes are practically continuous and so that a very small flywheel can be used.

CHAPTER XII

USEFUL INFORMATION

More About Pattern Making—Alloys and their properties: Red Brass; Standard Brass; Pewter; Fusible Alloy; Mitis Metal—Properties of some Useful Metals: Cast Iron; Soft Steel; Copper; Brass.

About Pattern Making.—In Chapter IV I told you a little about making patterns but there are some other things you should know in order to get good castings from them.

Pattern making is a very particular job and, if the pattern is at all complicated, it requires a good deal of *knowing how* in order to make the casting *draw* from the mold. The main thing in making simple patterns is to have the joints neat and close fitting.

If you should glue parts of the pattern together or get any grease on it, clean it off well before shellacking or the sand will stick to it and a poor casting will result. As I mentioned in Chapter IV the pattern must taper off slightly to permit it to be easily drawn from the sand.

In allowing for the shrinkage of the metal the casting is made of when it cools off the size of it must be taken into consideration. As iron shrinks 1/10 inch in a foot, brass 1/8 inch in a foot and steel and aluminum about 1/4 inch in a foot you can make your patterns accordingly.[1]

If the castings are to be left rough, they do not need to be made as large as when they are to be finished up either by hand or in a machine. For iron castings an allowance of 1/8 inch to the foot is usually made for the outside parts to be machined and 3/16 inch for the inside parts, while for brass castings 1/16 inch for the outside and 1/8 for the inside surfaces is enough for fair sized castings.

If a casting is to have a hole in it, as for instance, a cylinder, the pattern can be made solid and a *print,* that is a wooden plug the size and shape of the hole to be made is fixed to the ends where the hole is to be formed.

These prints must be marked as such when you send the pattern to the foundry so that the

---

[1] You can buy a *shrinkage rule* for $1.50 of Hammacher, Schlemmer and Co., Fourth Ave. and 13th Street, New York City. A shrinkage rule is graduated to allow for the shrinkage of different metals.

moulder will know they are intended for a hole and not solid ends.

Alloys and Their Properties.—An alloy is a metal made by melting two or more metals and mixing them together. Alloys take on properties that are entirely different from the metals they are made of.

Gun Metal.—This alloy has a very fine grain; it is yellow gray in color and in time past was used for gun castings. It is sometimes used now for high speed bearings and it makes very pretty model engine castings. It is composed of

      90 per cent. of *copper* and
      10 per cent. of *tin.*

Red Brass.—This alloy is very tough and is largely used for engine work. It is made of

      90 per cent. of *copper* and
      10 per cent. of *zinc.*

Standard Brass.—This is a brass alloy that makes good castings and works well. It is formed of

      66.6 per cent. of *copper* and
      33.3 per cent. of *zinc.*

Pewter.—This old-time alloy was formerly used for making plates, tea sets, etc. It is easy to work, is fairly hard and has a low melting point. It is composed of

>80 per cent. of *tin* and
>20 per cent. of *lead*.

Fusible Alloy.—It is also called *Rose's metal* because it was discovered by Valentine Rose; it dates back to 1772. It melts below the boiling point of water. It is formed of

>50 per cent. of *bismuth*
>25 per cent. of *tin* and
>25 per cent. of *lead*.

Mitis Metal.—This alloy melts and flows easily, has all the properties of the best forged iron and makes good castings. It is made of *Swedish iron* with $\frac{1}{700}$ to $\frac{1}{2000}$ part by weight of *aluminum*.

Properties of Some Useful Metals.—*Cast Iron.*—There are two kinds of cast iron and these are, (1) *white* and (2) *gray*. You can easily tell them apart for white cast iron is smooth and white, while gray cast iron has a grayish rough surface. Always have your castings made of gray cast iron because it is much softer and easier to work than white cast iron.

Soft Steel.—The *best* kind of mild, or soft steel has a sort of a fine bluish white color. The kind that is generally used for making machine parts is *cold rolled bright Bessemer* steel, but there are softer grades on the market.

Bessemer steel is easy working and can be filed, drilled, turned and threaded; in working it always use plenty of machine oil on it. It can also be bent cold and makes good forgings.

Copper.—Copper of good grade has a smooth, fine-grained surface, while the poorer grades are usually pitted and rough.

It is exceedingly *malleable,* which means that it can be hammered or rolled without breaking or cracking, and *ductile,* that is it can be drawn out into wire. It can also be forged at a low red heat, but it must not be heated anywhere near its melting point or it becomes very brittle.

Copper is not an easy metal to drill, file or turn, but it is easier worked if turpentine or soapy water is used to lubricate it. Its toughness and malleability makes it a very serviceable metal for model boilers.

Brass.—This alloy is quite malleable and ductile and makes very fine castings. You can saw, turn, file, thread and solder it easily and you do not need to lubricate it when performing these operations.

APPENDICES

APPENDIX A

Fluxes.—If metals which are to be soldered together are not clean the solder will not stick. To clean metals for soldering a *flux* must be used and fluxes of different kinds are needed for different metals.

Muriatic Acid Solution.—Muriatic acid, or hydrochloric acid to give it its right name, mixed with twice its quantity of water and in which zinc clippings are dissolved makes a soldering fluid that will solder tin, brass, copper, iron and steel, and nearly all other metals except aluminum.

When used on iron and steel wash it off very clean and dry well or it will rust the work. You can make this soldering fluid by buying 5 cents' worth of zinc chloride which you can get at any drug store, with an equal quantity of water.

Sal Ammoniac.—These salts are a good flux for soldering copper.

Tallow, Venice Turpentine and *Gallipoli Oil.* All of these are useful as fluxes for soldering lead and pewter and nearly all metals that melt at a low temperature.

Resin.—Strictly speaking this is not a flux since it does not clean off the grease and oxide but for soldering new tinware it can't be beat though you must scrape the tin clean first.

APPENDIX B

Model Engine Castings.—*Horizontal Steam Engine* Complete sets of castings for a horizontal engine can be bought in brass and iron at follows:

malleability makes it a very serviceable metal for model boilers.

Brass.—This alloy is quite malleable and ductile and makes very fine castings. You can saw, turn, file, thread and solder it easily and you do not need to lubricate it when performing these operations.

APPENDICES

APPENDIX A

Fluxes.—If metals which are to be soldered together are not clean the solder will not stick. To clean metals for soldering a *flux* must be used and fluxes of different kinds are needed for different metals.

Muriatic Acid Solution.—Muriatic acid, or hydrochloric acid to give it its right name, mixed with twice its quantity of water and in which zinc clippings are dissolved makes a soldering fluid that will solder tin, brass, copper, iron and steel, and nearly all other metals except aluminum.

When used on iron and steel wash it off very clean and dry well or it will rust the work. You can make this soldering fluid by buying 5 cents' worth of *zinc chloride* which you can get at any drug store, with an equal quantity of water.

Sal Ammoniac.—These salts are a good flux for soldering copper.

Tallow, Venice Turpentine and *Gallipoli Oil.*—All of these are useful as fluxes for soldering lead and pewter and nearly all metals that melt at a low temperature.

Resin.—Strictly speaking this is not a flux since it does not clean off the grease and oxides but for soldering new tinware it can't be beat though you must scrape the tin clean first.

APPENDIX B

Model Engine Castings.—*Horizontal Steam Engine.*—Complete sets of castings for a horizontal engine can be bought in brass and iron as follows:

(0) A horizontal engine with a ⅞-inch bore and a 2-inch stroke, all brass, for $4.00.

(1) Ditto with a 1½-inch bore and a 3-inch stroke, all iron, for $4.00, or all brass for $10.00, and

(3) Ditto with a 2-inch bore and a 4-inch stroke, all iron for $10.00.

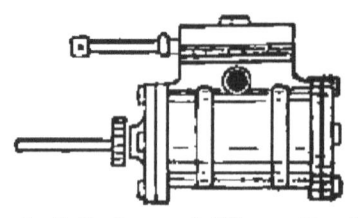

Fig. 81. A Cylinder and Piston Ready to Work

Parts for the Above Engines.—If you want to buy some of the parts finished and ready to work these can be had in iron and at the following prices:

Parts of Engine.	No. 0 Engine.	No. 1 Engine.	No. 2 Engine.
Cylinder and steam chest finished and in working order, as shown in Fig. 79	$10.00	$15.00	$20.00
Crankshaft and crank	2.50	3.50	5.00
Bearings for the shaft	2.00	2.50	3.50
Wrought iron connecting rod	1.50	3.50	5.00
Eccentric and straps	2.50	3.50	5.00
Flywheel, bored and faced	2.00	3.00	4.00

Castings of Cylinders.—Or you can get the *castings* for the cylinders in the rough and finish them yourself; they cost:

Diameter of Cylinder...	1 inch	1½ inch	2 inch
Price, bored	$1.50	$2.00	$4.00

The 1½- and 2-inch cylinders have the ports cast in them while the small cylinder which is made of brass must have the ports drilled in it and an extra charge of 75 cents is made for doing it.

Appendices

Castings for an Oscillating Engine.—A complete set of castings for making an oscillating cylinder engine of fairly decent size can be bought of model works at these prices:

No. 0 Engine with ¾-inch bore and 1½-inch stroke...... $1.75
No. 1 Engine with 1 -inch bore and 2 -inch stroke...... 2.50
No. 2 Engine with 1½-inch bore and 3 -inch stroke...... 3.00

When these engines are finished ready to run the No. 0 sells for $15.00; the No. 1, for $20.00, and the No. 2 for $30.00. The engine is shown in Fig. 80.

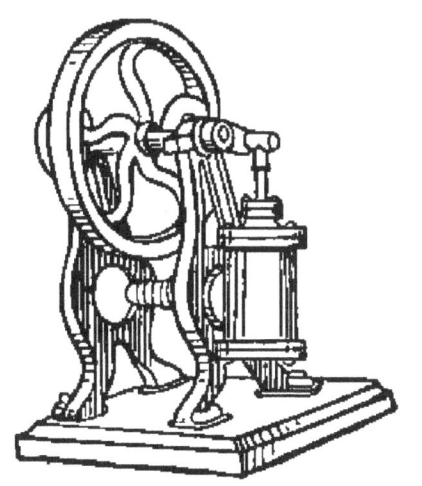

Fig. 82. An Oscillating Cylinder Engine

Castings for the Corliss Type of Engine.—An up-to-date horizontal engine of the Corliss type as shown in Fig. 81 is built just like a large engine, the design, construction and material being the same in every particular. The cylinder has a 1⅛-inch bore and a 2-inch stroke. Makers furnish them in three ways or classes as they call them, to wit:

Class A.—A complete set of castings, with all the ports cored in, including piston rod and valve stem. Price $4.50.

Class B.—Same as above with cylinder bored and faced, valves and steam chest seat planed and slide dressed. $5.50.

Class C.—Besides the above the balance wheel is bored, turned and polished, the stuffing boxes are fitted on and the eccentric is bored and turned. Price $6.50.

A finished engine of this kind costs $25.00.

Castings for a Half-Horse Power Corliss Engine.—This engine is made like the foregoing Corliss model,

Fig. 83. A Model Corliss Engine

but has a 1¾-inch bore cylinder and a 3-inch stroke, while the flywheel has a 2-inch face and is 10 inches in diameter.

The complete set of castings with the cylinder bored, steam chest and valve seat planed, crank shaft, piston and valve rods turned, $15.00, or you can buy the engine complete all ready to run for $50.00.

APPENDIX C

Steam Boilers for Model Engines.—*Vertical Tube Boilers.*—A vertical tube boiler large enough to run an engine of 1½-inch bore and a 3-inch stroke with all the fittings can be bought for about $30.00 and a size large enough for running a half horse power boiler for $50.00.

The smaller size boiler is 10 inches in diameter and 16 inches high and has 20 brass tubes ¾ inch in diameter and 10½ inches long. The shell is made of ⅛-inch wrought iron with lap welded seams, while the smoke-box and fire-box are of cast iron. The outside casting and smoke-stack are made of Russia iron.

The boiler is tested at 160 pounds cold water pressure and it is safe to run it at a steam pressure of 75 to 100 pounds.

Dimensions of Small Steam Boilers.—

Boiler Horse Power	Largest Base Diameter	Total Height	Fire-Box	No. of Tubes	Material of Tubes	Pounds Tested at	Water Line from Floor	Shp'g Weight in Pounds
¼	10-inch	21	5 x 7	20	Brass	250	14	90
½	16	31	6 x 9½	32	Brass	250	22	250
1½	21	41	9 x 13	51	Steel	200	27	425

These boilers are made to burn gas, gasoline, kerosene, hard coal, charcoal, alcohol or wood. If you are interested in a boiler of this kind write to the Lipp Electric and Machine Co., Paterson, N. J.

APPENDIX D

Other Castings for Engines and Boilers.—*Governors.* —A set of brass castings for a small governor can be purchased for 60 cents and a larger size for $1.00.

Force Pumps.—Three sizes of castings for force pumps are obtainable and these are

No. 1—¼-inch bore and 1¼-inch stroke.................. $ .50
No. 2—⅜-inch bore and 3 -inch stroke..................   .75
No. 3—½-inch bore and 4 -inch stroke.................. 1.00
The No. 1 force pump, finished, costs.................. 3.50

The No. 2 force pump, finished, costs.................... $4.50
The No. 3 force pump, finished, costs.................... 5.50

Safety Valves.—The castings for a small safety valve with a ¼-inch steam passage and a 5-inch lever costs 25 cents while a ⅜-inch safety valve with a ⅜-inch steam passage and a 6-inch lever sells for 35 cents.

Note.—The prices listed above are those quoted by the manufacturers at the time this book was written. Since these are war-times the prices are subject to change but I here give them so that you can at least judge the relative value of the different parts.

Where to Buy Engine Fittings.—Write for catalogues and price-lists to Arthur H. Wightman, 132 Milk Street, Boston, Mass.; The Chicago Model Works, Madison Street, Chicago, Ill.; Spon and Chamberlin, 123 Liberty Street, New York City, and the Weeden Toy Engine Company, New Bedford, Mass.

APPENDIX E

Speed Indicators.—*A Simple Way to Find the Speed of a Crankshaft.*—To find the speed that a shaft or a wheel is turning is simple if you have a *speed indicator*. It consists of a spindle on which threads are cut forming a *worm gear* and this meshes with a gear to which the dial is fixed.

Fig. 84. A Speed Indicator

By pressing the pointed end of the speed indicator against the center of the wheel or shaft the spindle turns the dial around and the number of revolutions can then be read off by timing it with a watch.

INDEX

Air pressure on piston, 6
Airplane engine, 23
Alcohol lamp, easily made, 29
Alloys, 243
Appendices, 247
Atlantic type of locomotive, 138
Atmosphere defined, 189
Atmospheres, steam measured in, 189
Atmospheric engine, 6
Automobile engine, 23
Automobile, steam, 16
Auxiliary parts of a horizontal steam engine, 88

Babbitt metal, 235
Ball governor, 120
Bearings for shaft of model turbine, 40
Bed plate for a horizontal steam engine, 70
Bed Plate, mounting engine on the, 86
Bed, setting engine on its, 86
Beighton, Henry, 9
Bell frame for locomotive bell, 142
Bell for a locomotive, 142
Beveled gears, 229
Blades, action of steam on turbine wheel, 43
Blades, making steam turbine, 33
Boil, how water is made to, 185
Boiler, a $\frac{1}{12}$ h. p. vertical tube, A simple iron, 92

Boiler—*continued*
Back tube sheet for locomotive, 147
To calculate heating surface of a, 197
Connections, 105
Crown sheet for a locomotive, 147
A good copper, 102
Firebox of, 97
Firebox sheet for, 95
Fittings for, 106
Fittings for a locomotive, 150
Front tube sheet for locomotive, 145
Furnaces, burners for, 99
Horsepower, 196
How to calculate the size of a, 196
How to test a, 107
Locomotive, 140
Safe way to operate a, 109
Safety valve for, 116
Seamless copper tube for, 95
See steam boiler
Shell of locomotive, 141
Smoke box, 97, 104
Smoke box sheet for, 95
Smokestack for a locomotive, 142
Steam dome for a locomotive, 142
Steam gauge for, 130
Steam in the, 189
Steam injector for, 126
Tubular, 140
Tubes for locomotive, 148

Index

Boilers, dimensions for small, 251
  Making small, 91
  For model engines, 250
  Safety valves for, 116
  Vertical tube, 250
  Whistles for, 113
Boiling, what it means, 186
Boulton and Watt, 193
Book of electricity, 174
Bourdon steam gauge, 131
Branca's engine, 3
  Impulse turbine, 43
Brass, 246
Buffer beam of locomotive, 161
Bunsen burner, 226
Burners for boiler furnaces, 97
Burner, kerosene, 100

Cab for locomotive, 174
Calibrating the dial of a steam gauge, 135
Caloric engine, 20, 199
Cardboard model of locomotive, 154
Cast iron, 244
Casting an engine cylinder, 72
Castings for boilers and engines, 251
  For the Corliss engine, 249
  For force pumps, 251
Castings for engines, 247
  Finishing engine, 72
  For a model engine, 62
  For oscillating engine, 249
  For safety valves, 252
  A steam chest, 75
Cawley and Newcomen's engine, 7
Central flue boiler, 104
  Flue steam boiler, 12
Centrifugal force, 120
  Governor, 12, 120

Clermont, Fulton's paddle wheel steamboat, 15
Cogwheel locomotive, 16
Compressed air engine, 20
Compression stroke, 239
Condensing engines, 190
Condensing steam, 8
Connecting rod bearing for steam engine, 81
Connecting rod for a steam engine, 80
Connecting rods of locomotive, 166
Connections for boiler, 105
Copper, 245
Copper boiler for a small engine, 102
Corliss engine, 249
Counterbalance weight, 164
Coupling rods for drive wheels, 165
Cowcatcher for locomotive, 173
Cradle, engine truck of locomotive, 160
Crank for a horizontal steam engine, 70
Cross-head guide block for steam engine, 78
Cross-head guide for a horizontal steam engine, 70
  Of locomotive, 157
  For steam engine, 77
Crown sheet for locomotive boiler, 147
Cylinder castings, 248
  Casting an engine, 72
Cylinder heads for a horizontal steam engine, 68
Cylinder of a horizontal steam engine, 66
Cylinders of locomotive, 157

Dead centers of an engine, 90
De Laval's steam turbine, 18

Index

De Laval's steam turbine—*ctd.*
   Model, 30
Development of the steam boiler, 12
Drawing plans for an engine, 64
Drawing tools you need, 62
Driving wheels locomotive, 164
Ductile defined, 245

Ebullition, 185
Eccentric for a horizontal steam engine, 70
Eccentric rod for a steam engine, 84
Eccentric for a steam engine, 83
Electrolysis, 185
Energy, heat a form of, 185
   Of motion, 187
   At rest, 187
Engine, airplane, 23
   Assembling a crankshaft of a gas, 233
   Assembling a gas, 235
   Atmospheric, 6
   Automobile, 23
   A ⅛ h. p. gas, 214
   Bed of a gas, 235
   Branca's impulse, 18
   Branca's steam, 3
   Caloric, 20
   Cam and camshaft of a gas, 228
   Camshaft bearings of a gas, 227
   Castings, cylinders, 248
   Castings, finishing, 72
   Castings, horizontal, 247
   Castings, model, 247
   Castings for oscillating, 249
   Castings of type metal, 62
   Compressed air, 20
   Connecting pipe for a hot air, 205

Engine—*continued*
   Connecting rod of a gas, 230
   Connecting rod of a hot air, 202
   Crankshaft and crank for a hot air, 206
   Cylinder of a gas, 216
   Exhaust valve mechanism of a gas, 220
   Expansion cylinder of a hot air, 200
   Firebox for a hot air, 208
   The first, 1
   The first real, 10
   Fittings, where to buy, 252
   Flywheel and pulley of a gas, 236
   Four cycle gas, 23
   Fuel mixture for a gas, 238
   Hero's reaction, 18
   Hero invents the steam, 2
   Gas and gasoline, 21
   Hot air, 20, 199
   Works, how a gas, 238
   Works, how a hot air, 210
   Igniter for an, 223
   Inlet and exhaust valves of a gas, 218
   Making a crankshaft for, 231
   Making a hot air, 200
   Newcomen's and Cawley's steam, 7
   Oiling a gas, 236
   Otto gas, 240
   Papin's piston steam, 5
   Parts of a gas, 216
   Parts of a hot air, 200
   Parts, mould of, 72
   Piston of a gas, 230
   Piston rod of a hot air, 202
   Power connecting rod for a hot air, 204
   Power cylinder of a hot air, 203

Engine—*continued*
  Power piston for a hot air, 204
  Self-acting, 8
  Standards for a hot air, 205
  Steam, 24
  Steam in the, 190
  Timing gears of a gas, 229
  To operate a hot air, 210
  Transfer piston of a hot air, 201
  Truck frame, 160
  Walking beam, 7
  Watt's double acting, 11
  Watt's rotary, 9
Eolipile, Hero's, 2
Eolus, god of the wind, 2
Ericsson hot air engine, 20
Exhaust stroke, 238
Expansion of steam in turbine nozzle, 44

Fire box for boiler, 105
  Of boiler, 97
  Grate of, 98
  Sheet of boiler, 95
Fire tubes, 13
Firetube boiler locomotive, 16
Fitch's steamboats, 14
  Steam packet, 14
Fittings for boiler, 106
  For a locomotive boiler, 150
  For model engines, 110
  Where to buy engine, 252
Fluxes, 247
Flyball governor, 120
Flywheel of engine, 11
  For a horizontal steam engine, 70
  Momentum of, 215
  For a steam engine, 85
Force pumps, castings for, 251
Fuel mixture for gas engines, 238

Fulton's *Clermont*, paddle-wheel steamboat, 15
Fulton's first steamboat, 15
Furnace for oscillating cylinder steam engine boiler, 58
Fusible alloy, 244

Gallipoli oil, 247
Gas burners, 99
  Defined, 188
Gas Engine, 21
  A 1/8 h. p., 214
  Assembling the, 235
  Assembling the Crankshaft of a, 233
  Bed of a, 235
  Cam and camshaft of a, 228
  Camshaft bearings of a, 227
  Connecting rod of a, 230
  The cylinder of a, 216
  Exhaust valve mechanism of a, 220
  Flywheel and pulley for a, 236
  Four cycle, 23
  Fuel mixture for, 238
  How it works, 238
  Igniter for a, 223
  Inlet and exhaust valves, 218
  Invention of the, 21
  Lebon's, 22
  Lenoir's, 22
  Making a crankshaft for a, 231
  Oiling a, 236
  Otto's, 22, 240
  Parts of a, 216
  Piston of a, 230
  Timing gears of a, 229
Gasoline engine, 21
Gears, reduction, 18
  For turbine reduction, 41
Governor, centrifugal, 12
  For steam engine, 120

Index

Grate of firebox, 98
Gauge, steam, for boiler, 130
   Water, for boiler, 136
Gun metal, 243
Gyroscope, toy, 31

Hanger, saddle and pedestal for locomotive, 144
d'Hautefeuille gas engine, 21
Headlight for locomotive, 174, 274
   Of a locomotive, 154
Heat a form of energy, 185
   As kinetic energy, 187
   Latent, 187
   As potential energy, 187
   What it does to water, 184
Heating surface, 13
Hedley's locomotive, 16
Hero of Alexandria, 1
Hero's steam boiler, 12
   Reaction engine, 18
   Reaction turbine, 43
Horsepower, boiler, 196
   How to calculate, 194
   Defined, 193
   Indicated, 196
   And work, 192
Hot air engine, 20, 199
   Connecting pipe for a, 205
   Connecting rod of a, 202
   Crankshaft and crank for a, 206
   Expansion cylinder, 200
   Firebox for a, 208
   How it works, 210
   Making a, 200
   Parts of, 200
   Piston rod of a, 202
   Power connecting rod for a, 204
   Power cylinder of a, 203
   Power piston for a, 204
   Standards for, 205

Hot air engine—*continued*
   To operate the, 210
   Transfer piston of a, 201
Horizontal steam engine, 61
Hydrogen, 182
Hydrostatic paradox, 127

Igniter of a gas engine, 223
Impulse engine of Branca, 4, 17
   Turbine of Branca's, 43
Indicated horsepower, 196
Initial pressure of steam, 44
Injector, steam, 126
   Steam, how it works, 130
Invention of the locomotive, 16
Isometric perspective, 64

Keeping up with your motor car, 223
Kerosene burner, 100
Kinetic energy, heat as, 187

Lamp for oscillating steam engine boiler, 59
Latent heat, 187
Lebon's gas engine, 22
Lenoir's gas engine, 22
Lever safety valve, 116
Link valve gear, how it is made, 168
   How it works, 179
   Locomotive, 167
Liquid fuel burners, 99
Locomotive, Atlantic type of, 138
   Bell for a, 142
Locomotive boiler, 140
   Back tube sheet for, 147
   Front tube sheet, 145
   To calculate heating surface of a, 197
   Crown sheet for, 147
   Fittings for, 150
   Steam dome for a, 142
   Tubes for, 148

Locomotive, cab for, 174
  Cardboard model of a, 154
  Connecting rods for, 166
  Cylinders, steam chests and cross-head guides, 157
  Dome for a, 142
  Driving wheels, 164
  With firetube boiler, 16
  Headlight, 154
  Headlight for, 174
  Hedley's, 16
  How it works, 176
  Invention of the, 16
  Link valve gear for, 167
  Link valve gear, how it works, 179
  Murray's, 16
  Parts of a, 140
  Parts of the engine, 156
  Pilot or cowcatcher for, 173
  Saddle, pedestal and hanger, 144
  Sandbox for a, 142
  Scale drawings of, 139
  Setting frame on truck wheels, 163
  Shell of a boiler, 141
  Side on coupling rods, 165
  Smokestack for, 142
  Steam pipe and throttle valve, 146
  Stephenson's, 16
  Stephenson's *Rocket*, 16
  Tender, 175
  And tender, finishing, 176
  Trailing wheels for, 173
  Trevithick's, 16

Major chord defined, 150
Malleable defined, 245
Metals, useful, 244
Metal working tools, 63
Mica for gas engine igniters, 224
Miller and Symington's steamboat, 15
Mitis metal, 244
Model De Laval steam turbine, 30
Model steam turbine, 24
Model steam turbine, tools needed to make, 32
Molecules of matter, 185
Momentum of a flywheel, 215
Moulding engine parts, 72
Mounting the engine on the bed plate, 86
Mounting the wheel of model steam turbine, 41
Muriatic acid solution, 247
Murray's locomotive, 16

Newcomen and Cawley's engine, 7
Newcomen and Watt, 192
Nozzle works, how a steam turbine, 44
  Making a model steam turbine, 35

Oil for gas engine, 236
Oscillating cylinder steam engine, 45
  See Steam engine oscillating cylinder
Otto's gas engine, 22, 240
Otto, gas engine inventor, 240
Oxygen, 182

Pattern making, 241
Patterns for a horizontal steam engine, 65
Papin's piston engine, 5
Parsons' steam turbine, 18
Pedestal, saddle and hanger for locomotive, 144

Index

Pewter, 243
Pillow blocks for a horizontal steam engine, 70
  For a steam engine, 86
Pilot, or cowcatcher for locomotive, 173
Pipe and Fittings for engines, 110
Piston engine, Papin's, 5
Piston for a horizontal steam engine, 69
Piston and piston rod for steam engine, 76
Packet, Fitch's, 14
Pop safety valve, 118
Potential energy, heat as, 187
Potter's self-acting engine, 8
Pounds, steam measured in, 189
Power stroke, 239
Pressure of steam, initial, 44
Propeller driver steamboat, 15
Pump, steam force, 123

Radius of a circle defined, 149
Reaction engine of Hero, 3, 18
Reaction turbine of Hero, 43
Rectangular steam boilers, 12
Red brass, 243
Reduction gears, 18
  For model steam turbine, 41
Resin, 247
Rocker arm for steam engine, 79
Rocket, Stephenson's locomotive, 16
Rotary engine, Watt's, 9

Saddle, pedestal and hanger for locomotive, 144
Safety valves, 116
  Castings for, 252
Sal ammoniac, 247
Sandbox for a locomotive, 142
Scale drawings of locomotive, 139
Seamless copper tube for boiler, 95
Self-acting steam engine, 8
Setting the engine on its bed, 87
Setting locomotive frame on truck wheels, 163
Shell of a simple iron boiler, 92
Shrinkage rule, 242
Side on coupling rods for locomotive, 165
Slide valve for a horizontal steam engine, 69
  And slide valve stem for steam engine, 77
Smoke box for boiler, 104
  Of boiler, 97
  Sheet for boiler, 95
Smokestack of a locomotive boiler, 142
Speed of a crankshaft, how to find, 252
Speed indicators, 252
Spring safety valve, 118
Standard brass, 243
Steam automobile, 16
Steamboats, the first, 14
  Fulton's first, 15
  Miller and Symington's, 15
  Propeller driver, 15
Steam boiler, 24
Steam in the boiler, 189
Steam boiler, central flue, 12
  Development of, 12
  Furnace for oscillating cylinder engine, 58
  Hero's, 12
  Making a toy, 24
  For oscillating cylinder engine, 5
  Rectangular, 12
  Tubular, 14

Steam chest, casting a, 75
  For a horizontal steam engine, 69
  Of locomotive, 157
Steam, condensing, 8
Steam dome for a locomotive boiler, 142
Steam engine, a 1/24 h. p., 61
  Assembling an oscillating cylinder, 53
Steam engine boiler, lamp for oscillating cylinder, 59
Steam engine, condensing, 190
  Corliss, 249
  Dead centers of a, 90
  Drawing plans for a, 64
  Fittings for, 110
  Governor for, 120
  High pressure, 190
  How to calculate h. p. of your, 194
  How to run oscillating cylinder, 59
  How it works, 89
Steam engine works, how an oscillating cylinder, 59
Steam engine of locomotive, 156
Steam engine, an oscillating cylinder, 45
  Parts of an oscillating cylinder, 47
  Parts for a 1/24 h. p., 65
  See Engine
  A simple piston, 45
  Stopcocks and taps for, 112
  Throttle valve for, 120
  Tools you need to make a, 62
  Tools needed to make an oscillating cylinder, 46
  Where to buy materials for oscillating cylinder, 50

Steam in the engine, 190
  Expansive power of, 11
Steam force pumps, 123
Steam, getting up, 186
  The giant power, 182
Steam gauge, 189
  For boiler, 130
  Bourdon, 131
Steam, how it acts, 189
  How it is measured, 189
Steam injector for boiler, 126
Steam locomotive, see Locomotive
Steam power plant, parts of a, 25
Steam pressure, about, 188
Steamships driven by steam turbines, 19
Steam turbine, 5
  Bearings, model, 40
  Blades, how to form model, 33
  De Laval's, 18
  How to run the toy, 29
  Model, De Laval, 30
  Model, reduction gears for, 41
  Modern, 17
  Multiple, 19
  Nozzle, how it works, 44
  Nozzle, making a model, 35
  Of Parsons, 18
  Single wheel, 18
  On steamships, 19
  Speed of model, 41
  Two simple, 24
  Wheel case, model, 37
  Wheel, mounting the, 41
  Wheel, making a model, 32
Steam turbine works, how model, 43
Steam whistle, 113
Steam, what it is made of, 182
Steel, soft, 245

Index

Stephenson link reversing gear, 168
Stephenson's locomotive, 16
Stirling hot air engine, 20
Stopcocks and taps, 112
Suction stroke, 238
Swage and Peening tools, 96

Tallow, 247
Taps and stopcocks, 112
Temperature of water, 186
Template for turbine blades, 33
Tender of locomotive, 175
Testing a boiler, 107
Throttle lever for locomotive, 153
Throttle valve, 11, 120
  For locomotive boiler, 146
Timing gears of a gas engine, 229
Tools you need, 62
Toy gyroscope, 31
Toy paddlewheel engine, 24
Toy steam boiler, 24
Toy turbine, how to run the, 29
Toy turbine wheel, 27
Trailing wheels for locomotive, 173
Trevithick's locomotice, 16
Truck frame of locomotive, 160
Tubular boiler, 14, 140
Turbine, steam, see Steam turbine
Turbine wheel for toy turbine, 27
Type metal castings for an engine, 62

Vacuum chamber, 190
Vacuum in cylinder engines, 6
Valve, spring and lever safety, 116
Valve stem and bearing for steam engine, 80
Valve, throttle, 11
Vapor, defined, 188
Venice turpentine, 247
Vertical tube boiler, 91
Vertical tube boilers, 250
Vibrations of matter, 185

Walking beam engine, 7
Water, to decompose, 185
  Formula for, 183
Water gauge for boiler, 136
Water, how it is formed, 184
  How it is made to boil, 185
  Temperature of, 186
  What heat does to it, 184
  What it is made of, 182
Watt, 193
Watt's double acting engine, 11
Watt, James, 9
Watt and Newcomen, 192
Watt's rotary engine, 9
Wheel case for model steam turbine, 37
Whistle, steam, 113
Wood working tools, 63
Work and horsepower, 192
  Time and effort, 193
  Unit of, 193

Lightning Source UK Ltd.
Milton Keynes UK
UKOW07f0006170216

268531UK00007B/212/P